TRIZICS

Teach yourself TRIZ,
how to invent, innovate and solve
"impossible" technical problems
systematically.

Gordon Cameron

TRIZICS™

WWW.TRIZICS.COM

Printed by CreateSpace 2010.

TRIZ and its concepts were created by G. S. Altshuller. The materials in this book seek to explain TRIZ in a modified way relative to the mainstream methodology of Altshuller's traditional form. The modifications and additions were performed with the intention of simplification and clarification. All reasonable effort has been made to ensure the reliability and accuracy of the information relating to the general concepts of Altshuller or other contributors to the TRIZ body of work. The author and publisher do not accept responsibility for the validity of materials or consequences of their application. Reprinted material is quoted with permission, and sources are indicated.

ISBN: 1456319892
EAN-13 9781456319892
Library of Congress Card Number: 2010916788

Direct all enquiries to info@TRIZICS.com.

Foreword

I first learned TRIZ (usually pronounced trees) several years ago with a group of fellow engineers. The consensus of opinion was that TRIZ was very powerful for igniting the creative imagination to solve problems, but there was a key issue - we did not have a way of selecting which TRIZ tools to apply or know when to apply them. Instead of being a process, TRIZ was a mêlée of disparate tools.

Almost every problem we faced as engineers in a high volume manufacturing environment was reactive in nature - something went wrong, we had to find out the root cause and then create a solution. TRIZ did not seem to help with finding root cause. We already had simple logical standardized methods for problem solving and for finding root cause that were highly effective. Only later did it become clear that TRIZ was not designed for finding root cause but instead was for finding innovative solutions to problems for which the cause was already understood. Yet, I believed many of the TRIZ tools could be leveraged to be highly effective for problems in need of root cause analysis.

The tools of TRIZ were clearly very effective at prompting creative thinking, but I became aware that much of TRIZ would benefit from further clarification and explanation.

A process for applying TRIZ was needed - a roadmap that incorporates the logic of standard problem solving, leverages TRIZ tools for root cause analysis, and directs the user to the appropriate TRIZ tool to use during the problem solving process.

The "TRIZICS" process was created as practical guide to problem solving using TRIZ. The process does not replace TRIZ, it organizes TRIZ tools into a single practical six-step problem solving framework which can be used to address any technical problem, including those for which root cause has to be determined. Many of the TRIZ tools in this book are explained in a new and I believe clearer way. All tools are accompanied by simple algorithms and basic illustrative examples which demonstrate how to apply them.

Without the roadmap, algorithms and explanations of how the tools work, TRIZ is difficult to follow and apply. Hopefully, TRIZICS will make the reader's engagement of TRIZ more logical, systematic and practical than ever before, lead to many innovative and practical solutions, and enable the reader to boldly go where no one has gone before in expanding the frontiers of technical problem solving and innovation.

About the Author

Gordon Cameron earned his degree in Physics at the University of Glasgow, Scotland. He has 25 years engineering experience in both research and development and high volume manufacturing in the semiconductor industry working in the United Kingdom, Republic of Ireland and the United States. He has published many technical papers on plasma etching systems used in the manufacture of semiconductor devices. His main focus has been on technical problem solving and troubleshooting. During his 16 years with Intel Corporation, he taught numerous TRIZ classes to engineers and has successfully applied TRIZ to many technical problems. Elected President of the Arizona TRIZ Association in 2007 Gordon has pursued the further development and application of TRIZ as well as other approaches to creative problem solving.

Acknowledgements

Many thanks to those who introduced me to the tools of TRIZ - Sergei Ikovenko and Alex Lyubomirskiy and to Jeff Brueck and Mark Delarosa of Intel Corporation for our many discussions. To Mark Cullen, for his comradeship, and Dr. Robin Obenchain for thinking outside-the-box. Most of all to my wife Cathie and our children Sally and Charlie for their love and patience.

References

Altshuller, G.S. (1988) Creativity as an Exact Science - The Theory of the Solution of Inventive Problems, Gordon & Breach, New York.

Altshuller, G.S. (2000) The Innovation Algorithm – TRIZ, systematic innovation and technical creativity, Technical Innovation Center Inc.

Altshuller, G.S. (2002) 40 Principles – TRIZ Keys to Technical Innovation, Technical Innovation Center Inc.

Cameron, G., (2008) Standard Structured Problem Solving, internal communication Caledonian Consulting LLC.

Cameron, G., (2005) Dynamization Evolution of Dry Etch Tools in Semiconductor Device Fabrication, www.triz-journal.com.

Ikovenko, S., Lyubomirskiy, A., 2005 TRIZ professional instruction notes.

Kowalick, J.F., (1996) Use of Functional Analysis and Pruning, With TRIZ and ARIZ, to Solve "Impossible-to-Solve" Problems www.triz-journal.com.

Mann, D., Dewulf, S., Zlotin B., Zusman, A., (2003) Matrix 2003 – Updating the TRIZ Contradiction Matrix.

Marconi, J. (1998) ARIZ: The Algorithm for Inventive Problem Solving an Americanized Learning Framework. www.triz-journal.com.

Rantanen, K., Domb Ellen, (2002), Simplified TRIZ. CRC Press LLC, Boca Raton.

Savransky, S.D. (2000) Engineering of Creativity CRC Press, Boca Raton.

Terninko, J., Zusman, A., Zlotin, B. (1998) Systematic Innovation – An Introduction to TRIZ the Theory of Inventive Problem Solving, CRC Press LLC.

Terninko, J., Miller, J., Domb, E., (2000) The Seventy-Six Standard Solutions, with Examples. February, March, April, June, July issues of www.triz-journal.com.

Zlotin, B., Zusman, A., (1992) Problems of ARIZ Enhancement, www.ideationtriz.com, originally published in *Journal of TRIZ* 3, No. 1, 1992.

Zlotin, B., Zusman, A., (1992) ARIZ on the Move, www.ideationTRIZ.com, originally published in *Journal of TRIZ* 3, No. 1, 1992.

Table of Contents

Chapter 1

Introduction to TRIZICS Problem Solving and TRIZ

Contents
1. **Introduction and Overview**
    1.1. The need for Creative Problem Solving
    1.2. Psychological Inertia
    1.3. What is TRIZ?
    1.4. Brief History
    1.5. Key Findings from the Study of Patents
    1.6. The Five Levels of Inventiveness
2. **Problem Solving Roadmap (TRIZICS)**
    2.1. Why do we need a Problem Solving Roadmap?
    2.2. TRIZICS Roadmap
    2.3. TRIZICS Roadmap Steps
        Step 1: Identify Problem
        Step 2: Select the Problem Type
            o Two Classes of Creative Thinking Tools
                o Analytical Tools
                o Solutions Tools
            o Which Analytical Tool to Choose
            o The Four Types of Problem
        Step 3: Apply the Analytical Tools
        Step 4: Define the Specific Problem
        Step 5: Apply the Classical TRIZ Solutions Tools
            o Which Solution(s) Tools to Choose
        Step 6: Compile Specific Ideas/Solutions
3. **Introduction to Analytical Tools**
    3.1. S-curve Analysis
    3.2. Trends of Evolution (also a Solutions Tool)
    3.3. Cause-Effect Chain Analysis
    3.4. Ideality
    3.5. Nine Windows
    3.6. DTC Operator
    3.7. Functional Modeling and Trimming
    3.8. Subversion Analysis
    3.9. Root Cause Analysis
4. **Introduction to TRIZ Solutions Tools**
    4.1. Classical TRIZ Solutions Tools
        Four Classical TRIZ Tools
            o Scientific Effects (1)
            o Inventive Standards (2)
            o Technical Contradiction (3)
            o Physical Contradiction (4)
    4.2. Trends of Evolution as a Solutions Tool
5. **The TRIZICS Process - Brief Summary**

Chapter Summary: *We are not typically educated in how to think creatively. We generally rely on our own talents. TRIZ is a set of tools for systematic creative thinking designed to help solve problems and create new, innovative ideas. It works by freeing our thinking and by applying the knowledge that has been used previously to solve breakthrough problems through the study of hundreds of thousands of patents.*

TRIZ tools are not normally organized into a step-by-step process. Instead TRIZ is typically presented as a set of disparate tools that are selected by the problem solver depending on the problem and situation. Inexperienced TRIZ users find it difficult to decide which TRIZ tool to use and when to apply it. To address this issue, TRIZ tools have been classified into analytical and solutions tools then organized and integrated with non-TRIZ problem solving tools to form a single comprehensive and logical problem solving process as shown in the TRIZICS Roadmap below in Section 2.2.

In our roadmap, all technical problems start at the same place - at the beginning of Step 1, the user is not challenged with where to begin. The roadmap directs the problem solver to the appropriate tools to use as the problem solving process progresses. Problem solving is systematic and easily followed. It is organized into six stages:

1. *Identify the problem.*
2. *Select the type of problem.*
3. *Apply the analytical tools.*
4. *Define a specific problem.*
5. *Apply solutions tools*
6. *Compile ideas and implement solutions.*

Each step of the problem solving process is discussed in this chapter, and an introductory overview for each tool is provided. Subsequent chapters provide detailed explanations, examples and step-by-step instructions on how to apply each tool. On completion of this book, the user will be able to apply the process to routinely to create the breakthrough solutions that drive invention and innovation.

1. Introduction and Overview

1.1 The need for Creative Problem Solving

Competitive demands require quicker, more effective and innovative problem solving. Problem solvers are required to quickly provide solutions to increasingly complex problems, develop and design new and innovative products and processes - all while reducing research and development time and costs.

To solve creative problems that require creative "out-of-the-box" thinking, we have historically relied upon the creativity of "talented thinkers." This is a random process that is not systematically replicable and relies solely on individual talent - an approach limited by personal knowledge, expertise, inspiration and luck.

Engineers and scientists spend many years learning the facts and technical details of their discipline yet there is generally no specific training provided for creative thinking skills. Creative thinking is a critical skill required by engineers, scientists and inventors yet it is generally done by trial and error - the thinker creates an idea and determines if it will work – which is basically guessing. Not only is trial and error limited by personal knowledge, it can take many ideas and a long time, sometimes years to identify a good idea and thinking is constrained by what is known as *psychological inertia*.

1.2 Psychological Inertia

Psychological inertia (also known as mental inertia) occurs when we make assumptions, usually subconsciously about a problem, resources and solutions. We subconsciously impose our own restrictions; rules and assumptions. This process is often driven by age and experience. The more experience we have, the more likely we are to use it, which freezes our thinking in place. The following simple problems illustrate psychological inertia:

Example 1
How many months have 28 days?
Example 2
At what speed must a dog run not to hear any sound from a frying pan that is tied to its tail?
Example 3
There are sixty lit candles in a room, but 10 have blown out. How many candles remain?
Solutions are at the end of this chapter (page 34).

Routine causes of psychological inertia are:
- Having a fixed vision (or model) of the solution or root cause.
- False assumptions (trusting the data).
- Language that is a strong carrier of psychological inertia. Specific terminology carries psychological inertia.
- Experience, expertise and reliance upon previous results.
- Limited knowledge, hidden resources or mechanisms.
- Inflexibility (model worship; trying to prove a specific theory, stubbornness).
- Using the same strategy. Keep thinking the same way and you will continue to get the same result.
- Rushing to a solution - incomplete thinking.

1.3 What is TRIZ?

TRIZ "**Теория решения изобретательских задач**" (**Teoriya Resheniya Izobreatatelskikh Zadatch**) is the Theory of Inventive Problem Solving that was first developed in the late 1940's. It was not until the 1990's that TRIZ first emerged from the former Soviet Union when a few TRIZ experts emigrated to the West bringing TRIZ with them.

TRIZ is a set of tools for directing creative thinking so that innovative ideas are not left to creative inspiration by trial and error. Instead, new and innovative breakthrough ideas that solve difficult technical problems and create new inventions can be systematically derived.

TRIZ can be applied or adapted to be used in many problem situations. It can help to identify the root cause of difficult to solve problems, identify innovative solutions, invent new products, proactively eliminate problems or failure modes, create new products, and improve maintenance procedures. It can be applied in any technical problem solving situation where brainstorming or creative thinking is required. TRIZ is not for calculation or engineering optimization; it helps create new ideas by providing the tools for breakthrough thinking.

It should be noted that the tools of TRIZ are not only useful to engineers and scientists but may be used by anyone who wants to think creatively. Many children have been taught to use several of the basic TRIZ tools. Many TRIZ tools have also been adapted for use in business applications and could be used to develop new software applications.

TRIZ tools are effective for:
- Quickly and systematically creating breakthrough ideas that solve simple or difficult technical problems that otherwise might take months or years to solve.
- Identifying root cause of existing problems (using modified standard TRIZ tools).
- Inventing new technical systems or technical processes.
- Improving the effectiveness and efficiency of existing technical systems or processes.
- Predicting future development of technical systems allowing early patent submission and setting research and development direction.
- Identifying and preventing future failures.

Before proceeding let's clarify the title. TRIZ, the "Theory of Inventive Problem Solving" is a translation that fits nicely into the often used English acronym, TIPS, but this translation can be confusing.

4

The word "theory" has two main meanings in English; the first common meaning is "speculation or guessing," the second is "an exposition of the principles of a subject." Theory has the latter meaning with reference to TRIZ, TRIZ is not speculation; it is based on the analysis of data collected by studying patents to understand the principles of how creative solutions can be achieved. The principles revealed by this study were used to make tools for creative thinking that allow the user to systematically produce breakthrough solutions.

The title "Theory of Inventive Problem Solving" is ambiguous because it could be interpreted as 1) the theory of how to inventively solve problems, or 2) the theory of how to solve problems that are inventive. The correct meaning is the latter; the theory of how to solve problems that are inventive. In TRIZ an inventive problem has a very specific definition; *an inventive problem is a problem that forms a contradiction.* An inventive solution is defined as a solution that solves a contradiction. A contradiction is simply a dilemma for example - if I make a change, then "something positive" happens, but "something negative" also happens as a result. Forming and solving contradictions is a core part of TRIZ that we will be discussing in detail. The Theory of Inventive Problem Solving could be more specifically stated as "the theory of how to make breakthrough solutions by solving contradictions."

Outside the former Soviet Union, the primary tool for prompting creative thinking has traditionally been "brainstorming" invented by the creativity theorist Alex F. Osborne (1888-1966). Brainstorming is usually carried out in a group situation, each member of the group suggesting ideas in sequence without criticism in an attempt to gather inventive ideas. Brainstorming is basically trying to create ideas or solutions by trial and error. It usually produces a lot of ideas, but many that are worthless (empty variants).

Successful brainstorming is dependent upon, but is also limited by individuals' knowledge, experience, creative talent and psychological inertia. The more difficult the problem is to solve, the greater the number of variants it takes to find a solution. This "trial and error" approach may be effective in situations where there are a limited number of solutions. However, if many alternative solutions are possible then trial and error can lead to a lot of empty variants. For example, inventor Thomas Edison used dozens of people and many trials to complete his many inventions. It is estimated that it took more than 10,000 trials to invent the accumulator battery by trial and error.

Few of us have the time or resources to rely on trial and error. Instead of trial and error, what if we could direct our thinking to produce only strong ideas? Instead of random, undirected creative methods where it could take years for a breakthrough idea, we can use TRIZ. TRIZ tools direct creative thinking and often provide breakthrough solution in minutes that otherwise may take months or years.

5

TRIZ is a set of tools for directing creative thinking. Classical TRIZ tools are based on data of how others have solved inventive problems in the past by studying the patent database. There are TRIZ tools that were developed for creative thinking - to release psychological inertia, and TRIZ tools that were created for breaking problems down so they are easier to solve. Our TRIZICS process combines the TRIZ tools with others into a single problem solving process.

1.4 Brief History

In the 1940's, G. S. Altshuller (1926-1998) the creator of TRIZ, started work on understanding how inventive solutions are created. Altshuller studied patents rather than trying to learn inventiveness by studying the psychology of great inventors.

He likened the study of patents to the study of chess. If you want to know how chess is played well, then study the games played by the grandmasters. If you want to know how to solve problems creatively, then study the patent database. The information in the patent database contained details of the problems that inventors faced and their innovative solutions.

Altshuller and his colleagues initially reviewed around 200,000 patents. To date, over 3 million have been reviewed by the followers of TRIZ and the results have stayed essentially the same. They are summarized below.

1.5 Key Findings from the Study of Patents

- Problems and solutions can be classified into 5 levels. The "**Level of Inventiveness**" or "degree of difficulty" increasing from 1 through 5.
- Higher level solutions (Levels 2 and up) required solving an "**inventive problem,**" one which involves the breakthrough thinking that solves a **contradiction.**
- The same few **inventive principles** were used to solve inventive problems.
- Problems and solutions were repeated across different areas of industries and science. Problems in one field had often already been solved in another.
- Innovations often used **scientific effects** outside the field from where the original problem was found.
- **Trends of technical evolution** exist, these show how technical systems develop over time, and that system evolutionary development is highly repeatable and therefore predictable.

Altshuller and his colleagues developed the classical TRIZ **solutions** tools based on those findings. By learning of how problems are solved innovatively from studying the patent database, tools that direct us to breakthrough solutions were created. Instead of random guessing, using TRIZ tools we can derive creative solutions using those tools.

The classical TRIZ **solutions tools** derived from studying patents are not the only TRIZ tools. TRIZ also uses a number of **analytical tools** that were developed to help define and understand problems and to help release psychological inertia.

In this book, in addition to TRIZ tools we also use a number of non-TRIZ tools selected to support our problem solving process, see the TRIZICS Roadmap in Section 2.2 of this chapter and in Appendix 8.1. We include a Standard Structured Problem Solving step first, which is a method of thinking inside-the-box using existing expertise. Often our goal is not to identify a breakthrough solution; many problems can be solved by the clear definition and organization of the problem without having to engage the TRIZ creative thinking tools. Additional non-TRIZ problem analysis tools are added, as is a root cause algorithm. Many problems require root cause analysis yet TRIZ does not include a standard useful tool for this. We will discuss the TRIZICS Roadmap in detail below.

Let's discuss the first on our list of Altshuller's findings from the patent database (Levels of Inventiveness), and then we will discuss the TRIZICS Roadmap, the various TRIZ and non-TRIZ **analytical tools** and the classical TRIZ **solutions tools** created from the study of patents.

1.6 The Five Levels of Inventiveness

Of the 200,000 patents, Altshuller initially selected 40,000 as representative of inventive solutions. Altshuller classified these patents into five "Levels of Inventiveness" with "Level 1" being basic, routine improvements and "Level 5" being highly innovative patents that required new scientific discovery.

Note that "Level of Inventiveness" does not correlate to the level of effectiveness. The levels simply indicate how difficult a problem was to solve in terms of requiring more ingenuity or knowledge from external fields of knowledge.

The percentages shown for each level in the list below represent the proportion of the total number of the sample of patents Altshuller reviewed. Of course these percentages may vary slightly depending on the specific sample used but the sample is seen as statistically large enough in size to be a valid representation of the entire database. Altshuller included an estimation of the number of trials it might take to obtain a solution using trial and error, a low number of trials correlates to the solution being fairly easy to find by trial and error.

Level 1
32%, less than 10 trials

Obvious or Routine Solution Level 1 patents provided a solution well known within its field of industry and technology; it often involves a simple adjustment or basic optimization. TRIZ does not classify Level 1 as inventive, since it did not require breakthrough thinking. Level 1 is an improvement that does not solve a contradiction.

Example: if I use a narrow hull the ship is unstable. Solution: use a wider hull. This is a Level 1 solution. There is no opposition to the obvious improvement solution; there is no identified contradiction. Level 1 does not change the system substantially. The hull was simply widened.

Note that a contradiction would be formed if there was opposition to widening. We will discuss contradictions and the three types of contradiction in detail later, but some basic understanding is needed to appreciate the categorization of the Levels of Inventiveness.

The *key* initial contradiction type to form when trying to solve a problem using TRIZ is a technical contradiction. Technical contradictions are formed when there is an obstacle to making an improvement; there is a dilemma such that making the improvement is opposed. A technical contradiction is usually stated in the form of an "if- then-but" statement:

> **If** (state what change is made),
> **then** (state what good happens),
> **but** (state what bad happens).

For example:

- **if** I widen the hull, **then** the ship is more stable **but** the ship moves slower.

And the opposite conflicting side of the technical contradiction is:

- **if** I don't widen the hull **then** the ship moves quickly **but** it is unstable.

It is always important to state the opposite conflicting side of a technical contradiction. To solve this problem requires a hull that is wide to be stable but narrow to move fast. By forming the problem into a contradiction a more innovative breakthrough solution is targeted. Creating the contradiction sets us on the path to creative innovation. Instead of thinking of a compromise solution (using slightly wider hull that's a bit more stable and moves a bit more slowly) we are directed to solve the problem with no compromise.

8

Instead of choosing a trade-off between one desired feature and the other, we target both, and thus achieve a breakthrough solution. We do not choose between stability and speed, we form the problem to achieve both stability *and* speed.

Thinking by way of contradictions is a powerful tool. By defining a contradiction we identify a problem that if solved will produce an inventive solution – a breakthrough idea or invention. In this case, a breakthrough solution is, for example, moving to a bi-hulled ship which is fast and highly stable.

Note also that because a Level 1 solution is very basic - simply widening the hull - this does not mean that a very basic change cannot solve a higher level of problem. Very basic changes can solve difficult problems if they involve a breakthrough solution that solves a contradiction.

Level 2
45%, up to 100 trials

The solution (the patent idea) is not well known within the industry or technology. It doesn't use knowledge from other industry or technology but requires creative thinking to provide the solution. Level 2 solutions solve a contradiction but not a significant one.

Level 3
18%, up to 1000 trials

Significant improvements are made to an existing system. The solution resolves a contradiction using engineering knowledge from other industries or technology.

Example: An electric field is used to move boxes rather than rollers. Contradiction: **If** I push the boxes, **then** they move **but** the boxes wear out. Solution: magnetic levitation.

Level 4
4%, up to 10,000 trials

Solution uses **science** that is new to that industry or technology. The move to a new scientific effect or phenomenon eliminates the contradiction, usually involving a radical new principle of operation.

Example: A sniper needs a bigger and bigger lens to accurately hit his target. Solution: use a laser sight to provide accurate location. Contradiction: **if** I use a bigger lens, **then** accuracy increases **but** it is difficult to carry a large lens.

Level 5
Less than 1%, over 10 million trials

Solutions involve discoveries of new scientific phenomena or a new scientific discovery. Contradictions are solved by discovering and applying new scientific phenomena. Level 5 inventions can lead to the creation of many new inventions at Level 4, 3, 2 and 1 levels of inventive solution.

Example: the solid state transistor took many years of research. A Level 5 solution solved the contradiction: if more electric valves are used then there is more computational power but the space needed becomes very large. Or if fewer electric valves are used then the space needed is reduced but there will be less computational power. The solution used a new scientific discovery.

The objective of TRIZ is to provide thinking tools for creative problem solving. Level 1 solutions do not require any creative thinking tools. It has been argued that Level 1 solutions, which are obvious simple incremental changes, should not merit a patent certification. Level 5 requires scientific discovery, and may be assisted by TRIZ but is essentially dependent on scientific discovery rather than problem solving. TRIZ primarily is used to create Level 2 through 4 solutions.

2. Problem Solving Roadmap (TRIZICS)

2.1 Why do we need a Problem Solving Roadmap?

TRIZ tools are not typically organized into a step-by-step process. Instead TRIZ is usually presented as a set of disparate tools that are selected by the problem solver depending on the problem and situation. Inexperienced TRIZ users find it difficult to decide which TRIZ tools to use and when to apply them.

There *is* an algorithm that is part of TRIZ, it is called ARIZ. ARIZ is the Russian acronym for Algorithm Rezhenija Izobretatelskih Zadach known as the Algorithm for Solving Inventive Problems, ASIP in English. It is an algorithm made up of nine parts that combines several TRIZ "solutions tools" into a sequential process. Each part has steps and it contains over 80 individual steps. ARIZ does not form an overall problem solving process, it is a method for taking a specific problem with known root cause and formulating and solving it as a contradiction. It is classified as a solutions tool on our TRIZICS Roadmap. ARIZ is discussed in detail in the Appendix 8.9.

Neither TRIZ nor ARIZ forms a comprehensive problem solving process. There is no traditional standard TRIZ process that leads us from the beginning of a problem solving process to the end, that is - from problem definition to implementation and validation of the solution. In order to systematically apply

TRIZ, we must organize the tools into a sequential process that we can use repeatedly for all technical creative problems from start to finish.

Without a single roadmap framework, TRIZ is difficult to use. Following seminars or training classes, most people don't know how to effectively apply TRIZ. There are simply too many choices of tool. The user has no guide for which tool to apply for the various types of problems there are, or at which stage of the problem solving process to use them.

Classical TRIZ tools are often inappropriately used to analyze problems or find root cause. Often the user simply does not know where to begin. This leads newly trained TRIZ practitioners to quickly become discouraged and, unable to use it, TRIZ is abandoned as being too difficult.

It is therefore important and a key objective of this book to try to provide a practical, organized problem solving framework for the application of TRIZ tools - an organized process that can be used consistently for all technical problems - a single roadmap that leads us from problem to solution for all types of technical problem.

The process we will use is "TRIZICS" (see TRIZICS Roadmap below, Section 2.2 and in Appendix 8.1); it includes all the necessary practical steps for a problem solving process. All technical problems start at the same place - the beginning or the TRIZICS Roadmap.

The TRIZICS roadmap defines the problem, specifies the limitations of what changes are allowed, establishes the timeframe for a solution, lists and checks assumptions, defines the success criteria, the cost, resources, and the implementation plan, etc. The practicalities of a competent comprehensive process from problem statement to solution implementation are considered, and organized into six sequential steps that include the key components of a practical structured systematic problem solving process. We consider the TRIZICS Roadmap key for the successful application of TRIZ for systematic problem solving and innovation.

Note that as we discuss the TRIZICS Roadmap in this chapter, the reader is exposed to references to new and unfamiliar tools and how they operate. This information may seem premature because the information is first described in detail in later chapters and has not yet been introduced.

Our approach is for the reader to achieve an overall understanding of the six step TRIZICS process initially in this chapter, and after progressing through the subsequent chapters, return with an understanding of the details. As each new tool is learned, the reader can refer back to this initial chapter and develop a clear understanding of where to use the tool during the problem solving process and the type of problem it is used to solve.

2.2 TRIZICS Roadmap

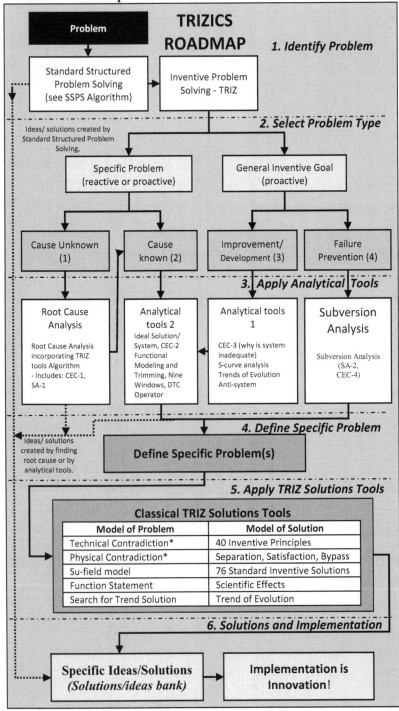

TRIZICS ROADMAP

Problem

1. Identify Problem

Standard Structured Problem Solving (see SSPS Algorithm) → Inventive Problem Solving - TRIZ

Ideas/ solutions created by Standard Structured Problem Solving,

2. Select Problem Type

Specific Problem (reactive or proactive)

General Inventive Goal (proactive)

Cause Unknown (1)

Cause known (2)

Improvement/ Development (3)

Failure Prevention (4)

3. Apply Analytical Tools

Root Cause Analysis

Root Cause Analysis incorporating TRIZ tools Algorithm - Includes: CEC-1, SA-1

Analytical tools 2

Ideal Solution/ System, CEC-2 Functional Modeling and Trimming, Nine Windows, DTC Operator

Analytical tools 1

CEC-3 (why is system inadequate) S-curve analysis Trends of Evolution Anti-system

Subversion Analysis

Subversion Analysis (SA-2, CEC-4)

4. Define Specific Problem

Ideas/ solutions created by finding root cause or by analytical tools.

Define Specific Problem(s)

5. Apply TRIZ Solutions Tools

Classical TRIZ Solutions Tools

Model of Problem	Model of Solution
Technical Contradiction*	40 Inventive Principles
Physical Contradiction*	Separation, Satisfaction, Bypass
Su-field model	76 Standard Inventive Solutions
Function Statement	Scientific Effects
Search for Trend Solution	Trend of Evolution

6. Solutions and Implementation

Specific Ideas/Solutions *(Solutions/ideas bank)* → **Implementation is Innovation!**

*Technical and physical contradictions may also be addressed using the advanced tools of ARIZ (see Appendix 8.9 ARIZ-85C) and the Contradiction Problem Solving Flowchart – Simplified ARIZ and Simplified ARIZ Algorithm (see Appendices 8.7 and 8.8 and Chapter 9).

There are several other tools used in TRIZ that are not included in the roadmap (Long Term Forecasting, Patent Circumvention, Feature Transfer etc.) those can be obtained from the published TRIZ body of work. If the user wishes to include them in the roadmap then simply decide which type of problems the tool is appropriate for and add it into one or more of the boxes in the "Apply Analytical Tools" step.

2.3 TRIZICS Roadmap Steps

Step 1: Identify Problem

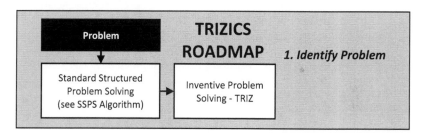

Our approach to applying TRIZ is to use it as a method for "out-of-the-box" creative thinking after we have first thought "inside-the-box." It is not a good idea to start to solve a problem by using TRIZ. Radical creative thinking should be used after logical thinking based on experience is applied.

When a technical problem is first encountered it is important to collect background information about the problem and to clearly define the objective or goal. What is the expected solution? What is the timeframe for a solution? What resources can be used? How and when will information be collected? How will the solution be implemented, etc? This basic information gathering is not a standard part of TRIZ problem solving methodology, but it is necessary if we are to provide a structured systematic process.

When using TRIZICS, we first apply a standard "in-the-box" problem solving method, allowing the user to use their existing knowledge, skills and creative thinking. We apply TRIZ and other creative thinking tools if the problem can't be solved or a more innovative solution is wanted.

Chapter 2 provides a specific method and template for Step 1 "in-the-box" Standard Structured Problem Solving.

Step 2: Select the Problem Type

Two Classes of Creative Thinking Tools

Our process is made up of a number of creative thinking tools. We can separate these into two classes, analytical tools and solutions tools. We will discuss them in a brief overview below after reviewing the steps in the TRIZICS Roadmap.

Analytical tools help us break the original target problem down to a level that is easier to solve or help us to see the problem differently. Usually, the analytical tools redefine the problem as a specific problem that if solved would solve our target problem or meet our inventive goal.

The solutions tools provide solutions to specific known problems (where known means we understand root cause). These are the classical TRIZ tools that lead us to creative ideas when a specific problem has been defined. We apply analytical tools first, then solutions tools. An overview of all the analytical tools and solutions tools is provided below in Section 3. Introduction to Analytical Tools and Section 4. Introduction to TRIZ Solutions Tools.

Analytical Tools
        1. S-curve Analysis
        2. Trends of Evolution (also a solutions tool)
        3. Cause-Effect Chain Analysis
        4. Ideal Solution/System
        5. Nine Windows
        6. DTC Operator
        7. Functional Modeling and Trimming
        8. Subversion Analysis
        9. Root Cause Analysis
        10. Anti-system

Solutions Tools
        1. Four Classical TRIZ Tools
            •   Scientific Effects (1)
            •   Inventive Standards (2)
            •   Contradictions
                ▪   Technical Contradiction (3)
                ▪   Physical Contradiction (4)
        2. Trends of Evolution
        3. ARIZ

Unlike the solutions tools, not all of the analytical tools are TRIZ tools. Some are borrowed to assist in the overall problem solving process.

Which Analytical Tool to Choose

The analytical tool to apply depends upon the type of problem. Not all of the analytical tools are appropriate to use for each problem situation. For example, if a pipe develops an intermittent leak (a problem with a specific goal, i.e. to stop the leak), then it would be inappropriate to apply a tool like S-curve analysis, which helps us identify how to direct the development of a technical system. For this problem, we need an analytical tool that first helps find the cause of the leak.

The appropriate analytical tool to use depends on the type of problem that you are trying to solve.

Our methodology chooses the analytical tools to use by problem type. We will define four types of problem and discuss how to choose the appropriate analytical tool. Note that the four problem types have no relation to the five Levels of Inventiveness.

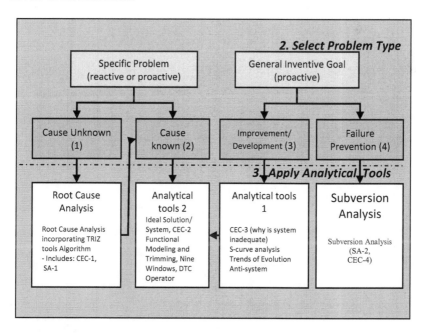

The Four Types of Problem

A specific problem may be reactive or proactive and the root cause may be known or unknown. A technical problem where the goal is general improvement is always proactive. General improvement involves improving, developing or inventing a technical system or technical process. We define preventing failures as a category of general proactive problem solving.

Four Types of Technical Problem		
Specific	Cause Unknown	Type 1
(reactive or proactive)	Cause Known	Type 2
General	General Improvement	Type 3
(proactive)	Failure Prevention	Type 4

Specific Problem (may be reactive or proactive)

Type 1: Solve a specific problem when the root cause is unknown.
Type 2: Solve a specific problem for which the root cause is known.

General Problem (proactive)

Type 3: Improve, develop, invent a technical system, or technical process.
Type 4: Prevent future failures for a technical system or technical process.

Examples of the four problem types:

Type 1: Eliminate intermittent leaking of a pipe in a gas supply system (specific, cause unknown).
Type 2: Eliminate the fracturing of a glass tube due to thermal expansion when it is heated (specific, cause known).
Type 3: Determine how to develop a motor car to gain market advantage. Invent a better floor cleaner. Improve the efficiency, repairability and quality of a preventative maintenance process (general inventive goal to improve a technical system or technical process).
Type 4: Eliminate the causes of failure for a metal electroplating process, a radio, roller coaster, a kettle (failure prevention).

Step 3: Apply the Analytical Tools

The analytical tools are classified into four groups:

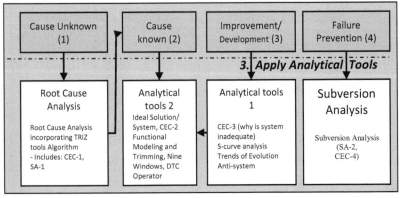

Root Cause Analysis Tools

Root cause analysis is used for a Type 1 problem i.e. a specific problem when the root cause is unknown. TRIZ is designed to be applied to technical problems for which the root cause is known and is not designed to identify root cause. Before applying TRIZ tools, you need to know the root cause of the problem. In our main roadmap, root cause is determined during Step 1, using Standard Structured Problem Solving. If root cause can't be found using standard methods, Chapter 11 of this book introduces a Standard Structured Problem Solving Incorporating TRIZ tools for Root Cause Analysis process, which shows how to leverage TRIZ tools that have been adapted to help find the root cause.

Analytical Tools 2

Analytical tools 2 are used for both Type 2 and 3 problems i.e. solving a specific problem for which the root cause is known and general improvement of a technical system or technical process. Inexperienced users are often confused by the fact that analytical tools can be used for both specific and general problems. The users often understand how to apply the tools to a general Type 3 problem but can't see how to apply the tools to a specific problem. We will explain how to apply the tools in the analytical tools 2 list to both types of problem in subsequent chapters.

Analytical Tools 1

Analytical tools 1 are used for Type 3 problems i.e. for the general improvement of a technical system or technical process. Note that trends of evolution can also be used as a solutions tool. We will discuss how to use it for both Type 2 and 3 problems. The ubiquitous Cause-Effect Chain analysis tool is used for each problem type. It is used differently depending on the problem type; we therefore separate it into four tools, CEC-1, 2, 3 and 4.

Subversion Analysis Tools

Subversion analysis (or failure prevention) is used for Type 4 problems, to predict future failures of a technical system or technical process.

Using the analytical tools alone often prompts ideas and solutions, because they help us to see the problem differently, and often inspire creative ideas without ever applying the solutions tools. Hence the TRIZICS Roadmap shows that we can add ideas to the solutions bank directly using analytical tools without using the solutions tools. If root cause is known, a solution may become clear without the need for TRIZ tools.

Step 4: Define the Specific Problem

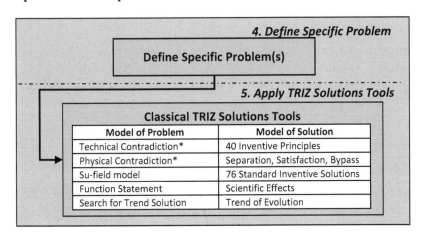

*Technical and physical contradictions may also be addressed using the advanced tools contained ARIZ (see Appendix 8.9 ARIZ-85C) and the Simplified ARIZ algorithm (see Appendices 8.7 and 8.8 and Chapter 9).

Applying the analytical tools often leads to a re-definition of the original problem and the definition of a number of different new specific problems to address. For example, the starting target problem may be how to stop a glass panel breaking, the specific problem we derive after analysis may be the how to stop vibration, or how to protect glass. These problems should be listed and kept track of in a "problems bank" where they can be prioritized.

Problems Bank		Priority
1		
2		
3		
4		
5		
etc		

After analysis, when the specific problem to be solved is defined, in order to apply the TRIZ solutions tools, we must re-state the problem in the form of one of the specific **Models of the Problem**. The specific problem must be stated as a technical contradiction, physical contradiction, Su-field model, function statement or a search for a trend of evolution. Below is a guide for which solutions tool to choose.

Step 5: Apply the Classical TRIZ Solutions Tools

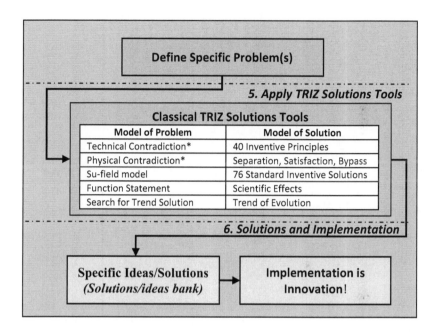

After the specific problem has been restated as one or more model of the problem, TRIZ identifies a model of the solution. This is a general solution analogous to the specific problem. It shows how others have solved a problem like yours in the past.

Classical TRIZ tools, ARIZ and trends of evolution of technical systems are the TRIZ solutions tools based on the study of patents that are used to try to find creative ideas for solving a specific known problem.

The TRIZ solutions tools for problem solving are:

1. Classical TRIZ (4 main tools)
  - Scientific Effects (see Chapter 7)
  - Inventive Standards (see Chapter 8)
Contradictions (two types) (see Chapter 9)
  - Technical Contradictions
  - Physical Contradictions
2. Trends of Evolution (see Chapter 4)
3. ARIZ we recommend using Simplified ARIZ (see Appendices 8.7 and 8.8 and Chapter 9) or Standard ARIZ (ARIZ-85C) which is discussed in the Appendix 8.9.

Which Solutions Tool(s) to Choose

To create the most ideas, users should expose the problem to as many TRIZ tools as possible. This is done by re-formulating or re-stating the problem in a way that is compatible with each solutions tool.

Technical contradictions, physical contradictions, Su-field models, function statements, inventive standards, scientific effects, etc. are currently unfamiliar terms to the reader. For now, to understand the process, it is important to understand that after a specific problem is defined by the analytical tools we must re-formulate it to be compatible with the TRIZ solutions tools.

An overview of the solutions tools is provided below in Section 4. It is a key skill in TRIZ to be able to state the problem in the various formats for applying the TRIZ tools.

The recommended guideline for which solutions tools to use is:

1. If the problem is simple and can be formulated as a basic interaction problem, form the problem into Su-field and function statements to apply the 76 inventive standards and search for scientific effects for solutions.

2. If a breakthrough is needed, form a technical contradiction (an "if-then-but" statement) and apply the basic tools for solving technical contradictions – the 40 inventive principles.

3. When there is a technical contradiction we can also form a physical contradiction and apply separation, satisfaction and bypass, assisted by scientific effects and a specific list of inventive standards.

4. If the problem is difficult to solve as a contradiction using basic tools, then apply the advanced tools for solving technical and physical contradictions. We took the key tools of ARIZ for solving technical and physical contradictions and named them the advanced tools. We created a flowchart to form and solve a contradiction which first applied the basic tools then the advanced tools (the tools of ARIZ). Our flowchart is named "Contradiction Problem Solving Flowchart - Simplified ARIZ" and is shown in Section 2.5 of Chapter 9 Part 1 and Appendix 8.7.

5. ARIZ is used to formulate your problem into a contradiction and then solve it. ARIZ is recommended for difficult problems and has a reputation of being difficult to use, we recommend using the Simplified ARIZ discussed above as a simpler alternative.

6. To apply trends of evolution as a solutions tool, simply review the each trend and determine if any inspire a solution.

Step 6: Compile Specific Ideas/Solutions

Because ideas can occur at any time, it is a good idea to maintain a "solutions bank." It is useful to maintain the information in the table shown on a chart on a wall during the problem solving process to track ideas.

Solutions Bank		
Source		Priority

Note that we state implementation is innovation. Innovation is the act of implementing a new idea or concept.

Now that the TRIZICS problem solving process has been described, let's get a slightly more in-depth review of the analytical tools and solutions tools we use in the TRIZ and the TRIZICS process. This will help familiarize and calibrate the reader with the sort of tools we will be discussing in detail later.

3. Introduction to Analytical Tools

In this chapter, analytical tools are introduced. Each is discussed in more detail in subsequent chapters.

3.1 S-curve Analysis (see Chapter 4)

Technical systems are created to provide one or more functions. Systems typically evolve by increasing their functionality and decreasing costs, where "costs" mean not only monetary value but includes harmful or negative factors. It is useful to define "ideality" as the ratio of functionality divided by cost. Systems therefore evolve by following a trend towards increasing ideality.

The life cycle of a technical system, component or subsystem can be described in terms of an S-shaped curve that develops in four main stages: infancy, growth, maturity and decline. The trend or "law" of increasing ideality is what drives a system up along the S-curve. By identifying the stage on the S-curve, specific actions can be taken. For example if a system or components' performance is determined to be at Stage 3 (maturity) then the S-curve predicts there will be no growth and decline may follow, functionality needs to be increased or costs reduced, it is a good idea to try to develop a new system. The typical indicators and recommended actions for each stage can be found in the TRIZ body of work and are not discussed in detail in this book.

S-curve analysis is not used for solving a specific problem. S-curve analysis is used to set direction for strategizing the development of a technical system.

3.2 Trends of Evolution (see Chapter 4)

From the study of patents, it was recognized that technical systems do not develop randomly but follow repeatable patterns or "trends of evolution." Knowing these trends allows users to predict the future development of a system.

Trends of evolution are mainly used for general, proactive problems (Type 3) where the user is trying to improve a system. The trends allow us to forecast the future technical changes in a system that can be exploited to maintain market leadership, obtain early patents or to create ideas to provide solutions to specific technical problems with known root cause (Type 2). Trends are not normally useful for helping find the root cause of a problem or for predicting future failure modes and are therefore not selected as analytical tools for Type 1 and Type 4 problems.

To use trends of evolution to improve a system (Type 3), we evaluate the system for its degree of development along the various trends. For example, we can examine how far evolved a system is for the trend of dynamization. If it has a rigid structure, then following the trend of dynamization, it should develop to become flexible in the future. For example, a screwdriver could develop to become more flexible, making it more functional, being able to be used in tight spaces and around corners.

For a specific problem (Type 2) we can review the trends of evolution. Reviewing the trends may inspire a solution. For example, we wish to increase the vibrational energy absorbed by an object. The trend of field dynamization indicates a development path of fields used by technical systems from a constant field, to gradient, variable, impulse resonance, then to an interference field. From this trend we create the idea to deliver the vibrations at the resonance frequency of the object to get it to absorb more energy.

Trend of Increasing Ideality
Technical systems drive towards becoming more ideal, they evolve to increase functionality and to decrease cost and harmful effects. A useful qualitative way of describing ideality is to use the relationship:

$$Ideality \sim \frac{Functionality}{Cost + Harmful\ Effects}$$

- Ideality (the value of the system) will increase if functionality increases.

- Ideality will increase if the payment factors, cost and harmful effects decrease.

The trend towards ideality is basic economics; it is the underlying trend that drives all other trends. Altshuller and his colleagues originally identified eight trends. To date, many have been added and more continue to be added. Trends such as the trend of increasing degree of trimming, the trend of decreasing human involvement, the trend of flow optimization, etc., may be found in the TRIZ body of work. We will discuss only the original eight trends created in the 1970's plus the trend of dynamization which was added in the 1980's.

1. The Trend of Increasing Completeness of a System
A simple system will evolve to a more highly developed one by accumulating four basic "functional blocks." It starts life with the operating part then adds another three parts usually in the following sequence: transmission, energy source and control system.

2. Trend of Energy Conductivity
In order for a system to operate, it needs to provide some energy to flow through all parts of the system. A system will not operate unless all parts work. The energy flow includes flow of information, substances, materials, objects, etc. A system will evolve to increase useful flows of energy and reduce harmful flows. A light bulb won't operate without the unhindered flow of electrical energy. A digital watch will not function unless it is driven by the flow of electrical power. A motor car requires energy to flow from the engine through the wheels in order to move. A freeway system won't operate when there is a blockage on the road.

3. Trend of Increasing Coordination or Trend of Harmonization
Systems will evolve to be more coordinated. For example, a driver's seat in a car developed to fit the shape of the driver.

4. Trend of Ideality was discussed above. All trends are driven by a trend towards becoming more ideal.

5. Trend of Uneven Development of System Components
The development of a system's parts typically grows unevenly – the more complex the system, the greater the number of irregularities.

6. Trend of Transition to the Supersystem
Over time, technical systems merge with the supersystem. As the system reaches its own limits of development in order to become more ideal it merges with systems from outside, incorporating external elements or external components.

6a. Trend of Dynamization
As a system develops it will trend towards an increase in the ability to change parameters in time (shape, temperature, size, etc. will "move"). It will become more dynamic and gain more freedom of movement.

7. Trend of Transition to Micro Level

The operating part or "tool" of a system begins as a macro level component or subsystem and develops towards the micro level. The operating part is the bristles on a paintbrush, the propeller of an airplane, the blade of a saw. Instead of macro objects, the work carried out by the operating part is performed by a new part or subsystem that operates at the micro level, which is the level of particles, atoms, molecules, electrons or particle fields.

8. Trend of Increasing Su-field Development

The evolution of technical systems will proceed in the direction of increasing Su-field development.

Su-field is an abbreviated term for substance-field interaction (Su-fields will be discussed in detail in Chapter 8). Typically this interaction is depicted as a triangle in which substance 2 (S2) acts on substance 1 (S1) via a field F. S2 and S1 can each be any "thing," such as an entire system, an object, or material, etc. but S2 and S1 cannot be a field. This complete three component interaction (Su-field) is also known as the minimum working "system."

The trend of increasing Su-field development is that incomplete, ineffective or harmful Su-fields strive to become effective complete Su-fields and that effective complete Su-fields will develop by increasing the dispersion of substances (the number and types of "things" will grow), the number of links between the elements will grow (more interactions by forming complex, double and chain Su-fields) and the responsiveness of the system will grow (the ability of the system to be controlled). Systems will also trend to more evolved field types. Mechanical fields evolve to acoustic to thermal to chemical and finally to electric and magnetic fields (sometimes the mnemonic MATCHEM is used to help remember this sequence).

3.3 Cause-Effect Chain Analysis (CEC, see Chapter 3)

CEC is a form of brainstorming tool that is preferred to standard brainstorming because thinking is logically directed. We have defined four CEC's. CEC is not a TRIZ tool but is a powerful analytical tool.

- CEC-1 is used to perform controlled brainstorming of the root cause.

- CEC-2 is used to perform controlled brainstorming of the root cause when root cause is believed to be known. We check we are working on the cause rather than a symptom.

24

- CEC-3 is used to try to identify problems for the improvement and development of a technical system or technical process.

- CEC-4 is used for failure anticipation, i.e. subversion analysis.

3.4 Ideality (see Chapter 4)

This tool can be used for a Type 2 or Type 3 problem. It uses the concept of ideality: the trend of increasing ideality is driven by basic economics. We want more for less. Systems with more value in a competitive economic system will survive and prosper.

A useful qualitative way of describing ideality is described above in the trend of increasing ideality.

- The *ideal system* is a system that has infinite functionality and zero cost and no harmful effects. To make a system more ideal we would make it more functional and/or perform its functions for less and eliminate harmful effects.

- The *ideal solution* is a solution that solves itself, for free.

3.5 Nine Windows (see Chapter 5)

This tool helps us think "outside-the-box." Instead of focusing our thinking on the present and system level, we also consider thinking and planning solutions in space and time - in the past and future at the subsystem and supersystem level. Nine Windows can be used for Type 2 and Type 3 problems.

The concept of the **anti-system** (creating ideas by considering the opposite function of a system, component, part, etc.), can also be used with Nine Windows although we mainly apply it to Type 3 problems, see 3.10 below.

3.6 DTC Operator (Dimension Time Cost Operator, see Chapter 5)

Like Nine Windows this tool is used for Type 2 and Type 3 problems. We consider how the problem changes if we think at extremes of Dimensions, Time (including speed) and Cost, which inspires creative ideas.

3.7 Functional Modeling and Trimming (see Chapter 6)

Functional modeling and trimming is normally applied to Type 2 and Type 3 problems, but can be useful for Type 1 and Type 4 problems also as a data gathering tool.

Functional modeling (also known as functional analysis) is a tool for capturing information about the way a technical system or technical process functions. We can use it to evaluate a technical system or process and reveal inherent problems or inefficiencies.

Functional modeling is a bit like drawing, but instead of making a representation of what a system or process physically looks like, it is a representation of the functions performed by components of a system or the steps of a process. Mapping what the components do (their functions) rather than what they look like releases psychological inertia that inhibits solving a problem or improving a system.

Trimming or pruning is a useful feature of functional modeling. When a model is created we can often see how to improve the system or process by removing, rather than adding, components or steps. Process functional modeling is a powerful auditing tool for identifying weaknesses in procedures.

3.8 Subversion Analysis (see Chapters 10 and 11)

Subversion analysis (SA-1) is used as a tool to find root cause by thinking of ways to cause the problem. Subversion analysis (SA-2) is used for failure analysis. We try to identify how to cause a problem and create a list of what can go wrong with a technical system or technical process. We then try to solve each problem proactively.

3.9 Root Cause Analysis (see Chapters 2 and 11)

Root cause analysis is first approached using Standard Structured Problem Solving (see Chapter 2). TRIZ was not originally developed to find root cause; it was designed to find solutions to problems where the cause was understood. Several of the TRIZ tools (and non-TRIZ tools) have been adapted to help find root cause. These are discussed in Chapter 11 Root Cause Analysis - Incorporating TRIZ tools and Appendices 8.5 and 8.6.

3.10 Anti-system: a tool for releasing psychological inertia by considering the anti-function or anti-action of a system, component, etc. The anti-system is sometimes used with Nine-Windows (see above Section 3.5).

4. Introduction to Solutions Tools

4.1 Classical TRIZ Solutions Tools

The solutions tools are classical TRIZ tools, ARIZ and the trends of evolution.

Solutions Tools

    1. Four Classical TRIZ Tools
- Scientific Effects (1)
- Inventive Standards (2)
- Contradictions
  - Technical Contradiction (3)
  - Physical Contradiction (4)

    2. ARIZ
    3. Trends of Evolution

There are four ways of solving a problem using classical TRIZ: using physical or scientific effects, using inventive standards, solving a technical contradiction or solving a physical contradiction. Also, there are trends of evolution which is not normally listed as one of the classical solutions tools but it can be used for finding specific solutions and so we include it as a solutions tool in our roadmap. There is also ARIZ which is not a single tool but an algorithm that uses several tools. We recommend using simplified ARIZ before standard ARIZ for ease of use.

Classical TRIZ tools operate by taking a specific problem and converting it to a generic model of the problem. By applying TRIZ tools, solutions based on the study of patents that solve this type of generic problem are identified. Prompted by the generic solution, users identify a specific solution for the original problem.

Take the following simple example: in the illustration below, which string should the mouse pull on to fetch the cheese, A, B or C?

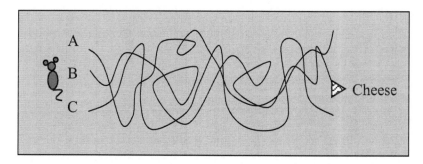

Instead of using trial and error, the simplest way to solve this problem is to start from the cheese and follow the string back to identify string A, B or C. By identifying the solution first, we can more easily solve the problem. Individual classical TRIZ solutions tools work in the same way. We find out what the solution "looks like" based on how others have solved a similar problem before, and then apply that model of the solution analogously to create our own specific breakthrough solution.

Classical TRIZ solutions tools use four basic steps; we start with a specific problem with known root cause.

my problem → a problem like mine → a solution to a problem like mine → my solution

This is often illustrated in the following way:

specific problem → model of the problem → model of the solution → specific solution

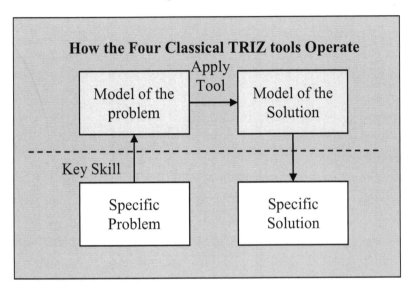

The model of the problem and model of the solution represent the individual classical TRIZ solutions tools created from the study of patents. Note that the flow in the diagram above matches the flow of our TRIZICS Roadmap between "define specific problem" and "solutions."

The key skill is to create a model of the problem from the actual specific problem. The table above illustrates the steps of how classical tools operate, (1) define specific problem, (2) state the problem in the format of one of the TRIZ tools (model the problem in a way the TRIZ tools will "understand"), (3) apply TRIZ solutions tools and obtain a generic solution based on the study of patents, and (4) create specific solutions based on analogy to the suggested general TRIZ solution.

So, after the specific problem has been defined (normally after applying the analytical tools), state the problem as a function statement, Su-field model, technical or physical contradiction, then apply the appropriate solutions tool (scientific effects, inventive standards, ways of solving a technical or physical contradiction) which creates a model of the solution, then prompted by that model of the solution, create a real specific solution.

Model of the Problem	Tool	Model of the Solution
Function Statement	Scientific Effects	Specific Scientific Effects
Su-field	Inventive Standards	Specific groups of Inventive Standards
Technical Contradiction	**Basic:** Contradiction Matrix and 40 Inventive Principles	Up to four specific Inventive Principles
	Advanced: ARIZ tools*	ARIZ tools for solving a Technical Contradiction
Physical Contradiction	**Basic:** Separation, Satisfaction, Bypass	Satisfy, Bypass, Separate (supported by table of Specific Inventive Principles to Physical Contradictions and research of Scientific Effects)
	Advanced: ARIZ tools*	ARIZ tools for solving a Physical Contradiction

* Note that we have classified ARIZ as a classical solutions tool for solving technical and physical contradiction at the advanced level. ARIZ is not one but is several tools combined.

In order to access the suggested solutions it is necessary to state the specific problem in the correct generic format - one that "makes sense" to each tool. Then find the generic solution.

Consider a multiplication table (tool) which provides the product of two numbers (solution). If the problem is; what is 7x9? We can look up the table, and find the solution, 63. Now consider another problem – how to stop a nail bending when it is struck with a hammer? We can't use the multiplication table. The multiplication table can only provide solutions if it is asked to solve the right sort of question. The problem must be in the format appropriate to the tool.

The key skill in applying the solutions tools is to be able to state the problem in the right format to apply the TRIZ tools. When the problem is in the right format, using TRIZ is easy. We are presented with a model of the solution - what the solution looks like, and we apply analogous thinking to create specific solutions. Much of the skill of applying TRIZ is problem reformulation. We will discuss how to formulate problems into the specific formats in later chapters as we discuss the solutions tools.

- To use **scientific effects**, we must reformulate our problem as a search for how to perform a physical effect, which we can state in the form of - *how to perform a function.*

- To use the **inventive standards** we must reformulate our problem as a search for ways of *how to improve an interaction* stating the problem as a Su-field.
- To use the **contradictions** tools we must state our problem as a contradiction. This is usually stated first as a technical contradiction in the form of an *"if-then-but"* statement: *if* I make this change *then* something gets better *but* something else gets worse.

TRIZ defines two types of contradiction that can be solved, technical and physical. We can try to solve contradictions using basic tools and more advanced tools if needed, the advanced tools are the tools contained in ARIZ. We recommend that the user first applies the basic tools to solve technical and physical contradictions then uses the advanced tools if more ideas are needed or the problem is not solved.

Technical Contradiction problem solving tools

Basic: Contradiction Matrix and the 40 inventive principles

Advanced: tools of ARIZ. (1.Use Su-field modeling and inventive standards to address the harmful action of each technical conflict. 2. Separate useful and harmful actions in time and/or space. 3. Use Smart Little People modeling to prompt ideas. 4. Make a resource list and form the ideal final result statement IFR-1).

Physical Contradiction problem solving tools

Basic: form physical contradictions by identifying antonym pairs (hot/cold, large/small, solid/liquid) and solve by satisfaction, separation or bypass.

Advanced: tools of ARIZ (perform resource analysis, and for each resource, identify antonym pairs to form IFR-2. Solve using separation, satisfaction or bypass. Also consider forming IRF-2 at the micro level).

In 1985 Altshuller moved to using Su-field modeling and the 76 inventive standards for solving technical contradictions instead of using the 40 inventive principles and Contradiction Matrix. We will use the Contradiction Matrix and the 40 inventive principles as our basic first tool because they complement the advanced tools (see Chapter 9 Part 1 Section 2.2).

The Contradiction Problem Solving Flowchart – Simplified ARIZ is discussed in Chapter 9 and shown in the Appendices 8.7 and 8.8. We will use this flowchart to solve contradictions.

We begin with the basic tools to solve a technical and physical contradiction and if needed we use the advanced tools to solve a technical and physical contradiction.

So the Contradiction Problem Solving Flowchart-Simplified ARIZ breaks solving an inventive problem into four phases A, B, C and D.

A. Form and Solve Technical Contradiction using Basic Tools.
B. Form and Solve the Physical Contradiction using Basic Tools.
C. Form and Solve the Technical Contradiction using Advanced Tools.
D. Form and Solve the Physical Contradiction using Advanced Tools.

The names and acronyms used in the flowchart (Appendix 8.7) will become clear upon reading Chapter 9.

4.2 Trends of Evolution as a Solutions Tool

To use trends of evolution as a solutions tool, we simply review the trends for inspiration as discussed above at the ends of Step 5 of the TRIZICS Roadmap and Section 2.3 above.

5. The TRIZICS Process - Brief Summary

Step 1 Identify Problem

Apply Standard Structured Problem Solving, ensure root cause is identified.

Step 2 Select the Problem Type

Decide if the problem is a specific problem or a general inventive problem or to prevent future failures.

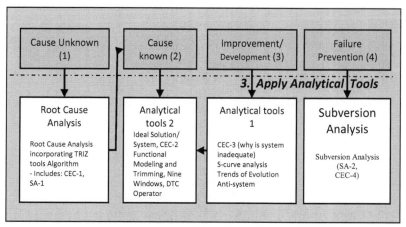

Step 3 Apply Analytical Tools

- S-curve analysis is used for a Type 3 problem and is used for planning future development of a technical system (product) or technical process. Main parameters of a system or process can be evaluated for their position on the S-curve and development strategies created.
- Trends of evolution are used for Type 3 problems. They are used to predict future technology developments and can be used to prompt future development ideas for new products or processes. Trends may also be used for Type 2 problems to prompt ideas for solving specific problems.
- CEC analysis is applied to all problem types. It is a logically directed brainstorming tool.
- The ideal solution/system is a powerful tool for release of psychological inertia. Ask - what is the ideal solution or system?
- Nine Windows helps us consider the past and future and the development of the subsystems and supersystems in our thinking and planning process.
- DTC Operator releases psychological inertia by considering extremes of dimensions, time, and cost.
- Functional modeling and trimming helps us gather information of how the components of a system function. It is a very powerful tool for releasing psychological inertia and prepares the problem situation for trimming.
- Subversion analysis helps us to find root cause - we think of ways to cause the problem, or it helps us to identify potential failure modes.

Step 4 Define Specific Problems

Formulate specific problems from the analysis. If the goal is to find a breakthrough solution, begin by forming a technical contradiction ("if-then-but") statement. If the problem is simple, form a function statement and Su-field and apply scientific effects and inventive standards.

Step 5 Apply TRIZ Solutions Tools

Apply Classical TRIZ Tools
- Function statement (search scientific effects for ways to perform a function)
- Su-field model (solve using inventive standards)
- Contradiction (two types)
  - Technical Contradiction
  - Physical Contradiction
  Use basic tools first for solving a contradiction, if needed, use advanced tools (see the Contradiction Problem Solving Flowchart - Simplified ARIZ Appendices 8.7 and 8.8 and Chapter 9) and Standard ARIZ - 85C if needed (see Appendix 8.9)
- Search through trends of evolution for a specific solution

The more tools the problem is exposed to, the more ideas will be created.

Step 6 Compile Specific Ideas/Solutions (use the solutions bank)

The TRIZICS Roadmap is accompanied by a template that the user may find useful. Roadmaps and templates are provided in the Appendices 8.1 and 8.2. By following the TRIZICS Roadmap, the user can start using TRIZ effectively (after reading the remaining chapters of this book). No matter what type of technical problem, the user should always start at the beginning of Step 1 of the TRIZICS roadmap.

The initial steps in the TRIZICS Roadmap instruct us to use the Standard Structured Problem Solving Flowchart (Appendix 8.3) and Template (Appendix 8.4). This is a process for thinking "inside-the-box." Frequently engineers and problem solvers can identify solutions by performing just the initial Standard Structured Problem Solving because the problem solvers thoughts become organized and the problem is more clearly defined. When it is clearly defined it becomes simpler to solve.

For a Type 1 problem, if the root cause remains unknown, having applied Standard Structured Problem Solving and the basic root cause analysis tools (propose model scenarios of the cause, Cause-Effect Chain, Fishbone Diagram or brainstorming), then we apply the Standard Structured Problem Solving incorporating TRIZ tools for Root Cause Analysis flowchart (Appendix 8.5) and template (Appendix 8.6). This includes additional information validation and adapted TRIZ tools for thinking "outside-the-box" to find root cause. Root cause must be known before proceeding further to use the TRIZ (analytical tools 2) and solutions tools. A Type 1 problem must be converted to Type 2.

When root cause is known (Type 2) or we have a Type 3 or 4 problem, TRIZ analytical and solutions tools are used to inspire the user to make creative solutions. The TRIZICS Roadmap proceeds by selecting the appropriate analytical tools which will produce a specific problem to solve.

We then reformulate the problem, using the guideline for which tool(s) to select (see "Which Solutions Tool(s) to Choose" Chapter 1 above). If a breakthrough inventive solution is sought, create or identify a contradiction and apply the basic tools for solving contradictions, that is, Sections A and B of the Simplified ARIZ Flowchart (Appendix 8.7) by following the Simplified ARIZ algorithm (Appendix 8.8). If needed apply the advanced tools for solving contradictions, Sections C and D by following the remaining parts of the Simplified ARIZ algorithm (Appendix 8.8). If preferred, for a difficult to solve contradiction problem, use ARIZ 85-C (Appendix 8.9). Applying the classical TRIZ tools, ARIZ and trends of evolution directs us to create a breakthrough solution.

At first, the TRIZICS Roadmap and supporting flowcharts (SSPS, SSPS with TRIZ Tools for Root Cause Analysis and Simplified ARIZ) look complicated, but they really are fairly straightforward to use. Imagine how complicated the instructions for tying a shoelace would appear when written down, it looks complex but really is simple after a little familiarization.

Solutions to psychological inertia examples above:

Example 1: All 12 months have 28 days.
Example 2: The answer is zero; if the dog is stopped there is no sound.
Example 3: There are still sixty candles.

Chapter 2

Standard Structured Problem Solving

Chapter Contents

1. Why we begin with Standard Structured Problem Solving
2. Standard Structured Problem Solving Algorithm and Flowchart
3. Standard Structured Problem Solving Details

Chapter Summary: *TRIZ does not have a tool for defining a problem or a process for gathering information that may be needed to understand all of the technical details necessary to solve a problem.*

The first part of our TRIZICS process is to apply the five step Standard Structured Problem Solving process is described in this chapter. This is used as an initial information gathering tool to define the problem, and as a method of trying to create "in-the-box" solutions before applying the tools of TRIZ.

Our problem solving always begins with this five step process before moving to TRIZ. Our method is to first gather information and clearly understand and define the problem before trying to solve it. Standard Structured Problem Solving directs us to first think inside-the-box then to think outside-the-box using TRIZ tools if needed.

Note that problems are treated differently depending on whether or not the root cause is known. If the cause is not known, it must be identified before trying to find solutions.

When the root cause is not known we try to identify it using one or more of three standard creative thinking tools used for root cause analysis: brainstorming, Fishbone Diagram and Cause-Effect Chain. Cause-Effect Chain (CEC) is discussed in more detail in Chapter 3.

When the cause is known, Cause-Effect Chain is used to check we are working on the right problem. When the problem has been correctly defined and cause is checked, the problem solver can propose solutions using their own creative talent or use brainstorming. Fishbone Diagrams are only recommended for root cause analysis.

If cause is known but a solution still can't be found or a more innovative solution is needed, we apply TRIZ. That is the traditional role of TRIZ, to find innovative breakthrough solutions to problems with known root cause, not to find root cause.

However, in Chapter 11 we show how many of the TRIZ tools can be modified to also be used to help find root cause. A root cause algorithm incorporating TRIZ is presented in Chapter 11 and Appendix 8.5. It is similar to the Standard Structured Problem Solving process that we will discuss in this Chapter but includes TRIZ tools adapted to help determine root cause.

In addition to the flowchart for Standard Structured Problem Solving, a template has been created to make it easy for the user to apply Standard Structured Problem Solving and is included in the Appendix 8.4. A flowchart and template for Standard Structured Problem Solving incorporating TRIZ Tools for Root Cause Analysis is included in Appendices 8.5 and 8.6.

1. Why we begin with Standard Structured Problem Solving

It is intuitive to try to solve technical problems by first thinking "inside-the – box," then thinking "outside-the-box." Many problems can be solved using routine standard problem solving based on existing available knowledge, skills and experience, without creating radical or innovative ideas. Divergent "outside-the-box" thinking is needed when a solution can't be found by standard means or when an improved innovative solution is wanted.

Our approach to problem solving will be to apply a Standard Structured Problem Solving method first to the problem then systematically apply TRIZ and other supporting analytical creative thinking tools. We do not recommend that the user begins to solve a problem using TRIZ tools.

We will treat standard problem solving as part of the "problem definition" stage of our problem solving roadmap since Standard Structured Problem Solving effectively gathers a lot of the technical background information that we need to understand the problem. It also ensures we have defined the root cause and understand the mechanism of the problem before trying to apply TRIZ.

If a problem is well understood and the problem solvers are stumped to find a solution or want a "better" solution, then using TRIZ tools id appropriate.

The Standard Structured Problem Solving described here is a basic generic problem solving method. It is not part of classical TRIZ and it is only a "suggested method." Any similar method may be used if it is disciplined and logically structured.

2. Standard Structured Problem Solving Algorithm and Flowchart

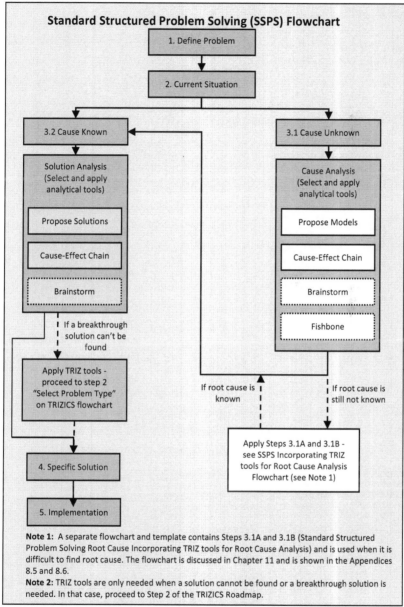

Standard Structured Problem Solving (SSPS) Flowchart

1. Define Problem

2. Current Situation

3.2 Cause Known

3.1 Cause Unknown

Solution Analysis (Select and apply analytical tools)

Propose Solutions

Cause-Effect Chain

Brainstorm

If a breakthrough solution can't be found

Apply TRIZ tools - proceed to step 2 "Select Problem Type" on TRIZICS flowchart

Cause Analysis (Select and apply analytical tools)

Propose Models

Cause-Effect Chain

Brainstorm

Fishbone

If root cause is known

If root cause is still not known

Apply Steps 3.1A and 3.1B - see SSPS Incorporating TRIZ tools for Root Cause Analysis Flowchart (see Note 1)

4. Specific Solution

5. Implementation

Note 1: A separate flowchart and template contains Steps 3.1A and 3.1B (Standard Structured Problem Solving Root Cause Incorporating TRIZ tools for Root Cause Analysis) and is used when it is difficult to find root cause. The flowchart is discussed in Chapter 11 and is shown in the Appendices 8.5 and 8.6.

Note 2: TRIZ tools are only needed when a solution cannot be found or a breakthrough solution is needed. In that case, proceed to Step 2 of the TRIZICS Roadmap.

Note that it is not necessary to use all of the analytical tools. It is recommended that the user first propose models/solutions then applies Cause-Effect Chain analysis. Use brainstorming and Fishbone tools only if needed in addition to Cause-Effect Chain analysis. The flowchart is supported by a template see Appendix 8.4.

Standard Structured Problem Solving Algorithm

Step 1: Define Problem
- Problem Statement
- Draw the Problem

Step 2: Current Situation
- Define Containment Plan (if needed)
- Problem Background and History
  o Problem Background Description
  o Data/Results
- Plans
- Resource Needs

Step 3 Analysis
- Root Cause Unknown
  o Investigate Root Cause
  o Key Conclusions
    ▪ Validate Data and Conclusion
  o Propose Models
  o List the Current Leading Model(s)
  o Apply Standard Tools for Creative Thinking
    ▪ Brainstorm
    ▪ Fishbone
    ▪ Cause-Effect Chain 1 (CEC-1)
  o If Root Cause still cannot be found, apply Algorithm for Root Cause Analysis incorporating TRIZ tools, (see Chapter 11)
- Root Cause Known
  o Propose Solutions
  o Brainstorm
  o Cause-Effect Chain 2 (CEC-2)
- If a solution or an innovative solution cannot be found, then apply TRIZ tools.

Step 4 Propose Solutions
Step 5 Implementation

3. Standard Structured Problem Solving Details.

Step 1: Define Problem

In this section the objective is to provide a clear problem statement. The current impact of the problem is described and quantified, the goal or expected outcome for the problem solver is described and limitations of what can and cannot be done or used to achieve the solution are clearly defined.

1.1 Problem Statement
State the problem or inventive goal. Communicate the essence of the problem in a concise form. What's wrong? Why is it a problem? What is the goal of solving the problem? Quantify specific details if possible. For example:

- A hot water pipe in the engine room of a ship forms leaks at an average rate of once per week, resulting in the loss of 20 hours production time and the loss of 50 parts, at a cost of $20,000 in lost revenue per month. The goal is to reduce the number of leaks to zero.

Draw the Problem
When drawing a problem, include a side and top view if possible. Include any pictures, photographs, engineering drawings, etc. The drawing does not have to be a work of art, but should contain the key components for releasing psychological inertia to aid thinking.

1.2 What is the name of the technical system or technical process that is the target of the problem in which the problem resides?

1.3 Describe the main useful function of the technical system or technical process (this can be a sub process or process step of interest).

1.4 What is the impact or cost of not solving the problem?
- Capture and quantify the gain of solving the problem as specifically as possible. Why are you solving this problem? Often, problems are pursued without validating whether there is a return on investment.

1.5 What are the success criteria, the definition of "problem solved"?
- Define what "solved" means. It's important to communicate with all stakeholders the definition of success to avoid confusion. Clarify whether the goal is a quantifiable measure of improvement or complete elimination of the problem. All stakeholders should be clear on what "problem solved" means.

1.6 What is the timeline for getting a solution?
- Define when a solution is needed or when specific milestones should be complete.

1.7 What are the limitations and requirements?
- Identify any restrictions or limits on resources involved in trying to find a solution. For example, project should take no more than 10 hours per week, can't include the use of unsafe chemicals, can't use the color blue, simple low tech solutions are preferred and so on.

Step 2: Current Situation

In this section we capture the background and history of the problem including the current status of the problem. We document the results of previous work or experiments that have been done or are in progress, and what is currently planned.

2.1 Containment of the problem. Is there a containment plan needed (quick fix or compromise, or "band aid" to mitigate the problems impact? If so, develop and implement a containment plan.

2.2 Problem Background and History
Problem Background Description:
- Describe the technical details of the problem, how the technical system or technical process normally operates. This may require a technical content expert.

- If relevant, list competing or similar systems or processes and describe their performance and operation, etc.

Data/Results:
- Summarize existing data and information. Describe when the problem was first observed. Identify if the problem is intermittent or continuous, etc. Gather all data analysis, results of any experiments, graphs, trend charts, tests for commonalities to specific tools resources or circumstances, etc.

2.3 Plans
Summarize the current planned actions, planned data analysis, planned experiments or experiments in progress and identify their purpose. If root cause is unknown, investigate root cause (see Step 3.1 below).

2.4 Resources Needed
Having reviewed the problem definition and background, list any specific resources that are estimated or may be needed to solve the problem.

Step 3: Analysis

During the analytical stage, the problem situation, background information and data are examined. Creative thinking tools are used to try to reveal possible causes and solutions. The tools to apply depend on the type of problem. For Standard Structured Problem Solving we distinguish between root cause known and unknown.

3.1 Root Cause Unknown

To solve a problem for which the root cause is unknown, it is almost always necessary to first identify the cause of the problem, otherwise a solution usually cannot be found. We must troubleshoot the root cause and identify the mechanism in order to create a solution.

Investigate Root Cause

Review the information gathered in Steps 1 and 2 and propose potential models of the cause. Capture them in the "Propose Preliminary Cause Models" Section 3.1.1 below. Identify any gaps in knowledge or information that may be collected or analyzed to help support or eliminate a cause model. Some key actions are:

- **Look for Commonalities**. Is the failure mode common only to a particular resource, machine, tool, person, batch of materials, time of day, time of year, particular environmental conditions, storage conditions, temperature, or specific combinations of resources or circumstances?

- **Look for Interactions**. Is the failure mode associated with specific combinations of resources, machines, tools, etc?

- **Look for Trends**. Did the failure occur gradually over time or usage due to an "aging effect," etc? What change(s) are coincident with the turn on or ageing effect?

- **Run Segmentation Experiments**. Perform experiments to try to isolate (segment) where, when or what particular circumstances are needed to create the problem.

Summarize Conclusions

Summarize the information gathered in Steps 1 and 2 and from the root cause investigation above. List key conclusions that can be drawn and how the data was validated.

Data Conclusion Table

	Data/Assumption	Conclusion	How data was validated
1	Pipes only leak after 3 months	Pipes are being slowly eroded	Analysis of recorded install dates.
2	Only acid pipes leak	Acid corrodes the pipes	Checked flows through all pipes
3	Only large pipes leak	Small pipes don't erode	Assumption
Etc			

Psychological inertia, erroneous data, wrong assumptions and false conclusions are often why root cause is not identified. It is important to keep an open mind and eliminate psychological inertia.

It is important to validate and challenge all data and assumptions. It is necessary to prove the data being used is valid and to ensure the correct conclusions are drawn.

Below is a list of frequent reasons root cause is not identified and recommended actions the problem solver should take to address those issues.

Frequent reasons root cause is not identified:

- **Experience**: too much experience can lead to psychological inertia and drive thinking in a "trained" direction closing off new ideas.
- **Fixed thinking techniques**: repeating the same steps and using the same methods leads to repeating the same result, creating the same ideas.
- **Group think**: over time, a set of individuals working on a project will tend to think the same way, believe the same conclusions and results. This group mindset leads to psychological inertia, as new members are introduced instead of pursuing new ideas provided by "a fresh pair of eyes," the group tries to assimilate new members to existing thinking.
- **Model worship:** a specific "favorite" model is pursued and alternatives are dropped.
- **False information/incorrect data/false assumptions**: this may be due to the way the data or information was collected. For example, incorrect calibration of a measurement standards, or simply incorrect information or facts have been obtained or assumed.
- **False Conclusion**: for example, the sun rises every day in the east. False conclusion - the sun revolves around the earth.
- **Hidden resources**: the problem is caused by contaminants, or secondary or derived resources.
- **Hidden mechanism**: mechanism may be a new or unusual phenomenon or be an effect outside the problem solvers field of engineering or science.
- **There is no actual problem**: the result is "normal" simply a "rarity" or "outlier" of the normal distribution.
- **There is more than one problem** and therefore more than one cause.
- **Insufficient technical knowledge**: this is rarely the reason for a problem's root cause not being identified. Normally such gaps in knowledge are quickly closed and problems solved.

Actions to address common issues that impede root cause determination

- Have new people check all data and information to provide fresh thinking.
- Determine whether the conclusions can be wrong (be highly critical of all conclusions).

- Check the information is indisputable; assign a specific person (owner) responsible for checking the data.
- Physically check and visually witness information or data rather than accepting validation from others.
- Always challenge calibration methods.
- Determine what potentially hidden or secondary resources might be present and how they could cause the problem.
- Describe a new or unusual mechanism that would have to exist to cause the problem.
- Demonstrate the problem is not simply an outlier (a rare but expected event and therefore not a "problem" at all).

Independent validation of each piece of data, assumption and conclusion is needed. It is useful to list all assumptions and conclusions and challenge each in turn. Re-checking the information, using different personnel is necessary to avoid psychological inertia.

State the Leading Model of the Root Cause
It is useful to capture the current leading model of the root cause (the model considered to be most likely) as a communication tool. If there are several good models then they may all be documented.

	Leading Model(s)
1	The composition of pipes changed by manufacturer causing increased erosion.
2	The temperature increased due to a faulty thermocouple.
3	
Etc	

When creating a model or scenario of how the problem is caused, it is useful also to consider Ockham's razor. Ockham's razor states the most likely cause or explanation is often the simplest. This doesn't rule out complex causes, but it does help prioritize which to validate.

3.1.1 Propose Preliminary Cause Models

When there is a reactive problem with unknown cause, the first instinct is to understand the cause of the problem. It is useful to capture these initial ideas. Any idea or model of the cause must be consistent with the data.

Let's say we have a specific problem, our bicycle tire is flat. What is the root cause?
Models:

	All Models
A	The tire got punctured when cycling over some glass.
B	The valve became loose after going over some rough ground.
C	The tire was old and worn.
D	Etc

Standard Tools for Creative Thinking
It is helpful to apply standard tools for creative thinking to try to create ideas. Later we will learn how to apply TRIZ tools to help find root cause.

3.1.2 Brainstorming

Brainstorming is normally done as a group activity. It is a process for gathering creative ideas to solve a problem. The target problem is stated, potential causes or solutions are identified by moving sequentially around the room to get ideas from individuals. All are encouraged to use their creative mind to generate ideas. Of key importance is that ideas are generated with no criticism to try to liberate the less logically derived ideas and to open the imagination. The most radical solutions are derived towards the end of such a session. When few ideas are being created, the most radical come to the fore and these are what we are trying to find.

3.1.3 Fishbone Diagram

Ideas are organized into similar categories using a diagram that resembles a fishbone. Also known as an Ishikawa diagram (after its creator *Kaoru Ishikawa* 1915-1989) or cause and effect diagram (not to be confused with Cause-Effect Chains). The cause may be easier to find when broken down into specific categories.

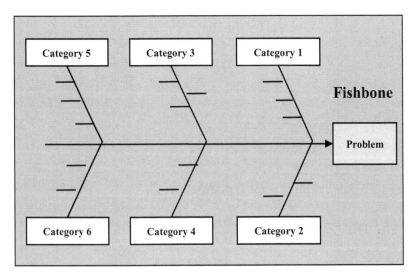

For example, there are 10 tanks used to chemically remove metal from an object. One specific tank needs to be cleaned more frequently than the others. Use a Fishbone Diagram to help find ideas for the root cause.

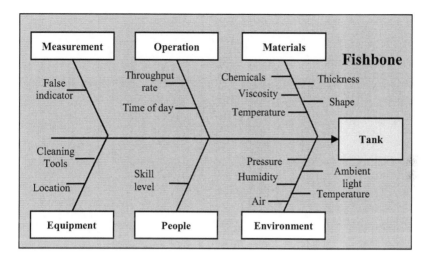

The factors identified (location, temperature, etc.) can now be considered as a potential root cause.

3.1.4 Cause-Effect Chain Analysis (CEC-1)

The recommended tool to use first for root cause analysis is Cause-Effect Chain analysis (CEC) because it applies a form of logically directed brainstorming. Cause-Effect Chain analysis is discussed in Chapter 3.

3.2 Root Cause Known

When the root cause is known, then the inventive challenge is to find a way to eliminate the cause or to provide the desired effect. For example:
- The pipe leaks because it is eroded from the inside by the flow of materials inside.
- The chair is made of iron and is therefore too heavy to carry.

We understand what the problem is; the difficulty is finding a solution. In order to find a solution, it is useful to simply propose ideas based on (in-the-box) knowledge and experience, then to apply brainstorming and Cause-Effect Chain analysis (CEC-2).

Propose Ideas (use imagination)
List ideas and solutions based on knowledge and experience (remember these will be subject to psychological inertia).

Brainstorm

Brainstorming is conducted in the same way as described in 3.1.2 above, the difference being here you try to find a solution not root cause.

Cause-Effect Chain Analysis (CEC-2)

Is conducted in the same way as CEC-1, described in 3.1.4, the difference being the specific problem cause is believed to be understood. The CEC is repeated to check our understanding and validate root cause. It is used to check we are working on the cause and not the symptoms. It is not necessary to repeat this second CEC-2 if CEC-1 has been validated.

If a solution cannot be found, or a breakthrough solution is needed, then apply TRIZ tools by engaging the TRIZICS Roadmap at Step 2 - Select Problem Type.

Step 4: Solutions

In this section we try to identify solutions to eliminate the root cause or make the inventive improvement by solving the problems identified during cause analysis. In the absence of TRIZ we "brainstorm" solutions.

Because ideas can occur at any time, it is a good idea to maintain a solutions bank. Also note the source of the idea. Was it personal creativity, Cause-Effect Chain, a specific TRIZ tool?

Solutions Bank		
Source		Priority

Step 5 Implementation

In this section we create a standard implementation plan that includes validation, monitoring, dissemination of the solutions. Document any follow up actions required, owners assigned, completion dates, success criteria of implementation, etc.

	Action	Priority	Projected Completion Date	Owner	Status
1					
2					
3					
Etc					

Standard Structured Problem Solving Template – see Appendix 8.4.

Chapter 3

Cause - Effect Chain Analysis

Chapter Contents

1. **CEC Application by Problem Type.**
   1.1. CEC-1:For Finding Root Cause (Type 1 problem)
   1.2. CEC-2:To Validate Root Cause (Type 2 problem)
   1.3. CEC-3:Proactive Improvement of a Technical System or Technical Process (Type 3 problem).
   1.4. CEC-4:Prevent Future Failures for a Technical System or Technical Process (Type 4 problem)

Chapter Summary: *We briefly discussed Cause-Effect Chain analysis (CEC) in Chapter 2. CEC analysis is best described as "logically directed brainstorming." It is a versatile tool and can be used in many problem situations. It works by asking the simplest of questions: "why?" It provides a method for us to progressively and logically think our way to the root of the problem.*

We discuss how to apply CEC's for each of the four problem types we defined for our problem solving roadmap. The way we use it is slightly different for each. For example, for a specific problem for which root cause is unknown, we start with a specific statement "the pipe is leaking – why?" For a general proactive problem we start with the general statement, "the technical system is inadequate" then ask "why?"

CEC Application by Problem Type

Cause-Effect chain analysis is a simple but very powerful analytical tool. It is best described as "logically directed brainstorming." It is not a "TRIZ tool" but is used as part of the overall problem solving process.

TRIZ itself is not effective for problem definition. TRIZ is useful after specific problems are defined and the cause is known. Cause-Effect Chain analysis provides an effective method for helping with problem definition, we use it as part of the Standard Structured Problem Solving process and as an analytical tool that can be applied to any type of problem.

Cause-Effect Chain analysis has therefore been integrated into our problem solving process (see TRIZICS Roadmap in Chapter 1 and Appendix 8.1).

Below is a section of the TRIZICS roadmap.

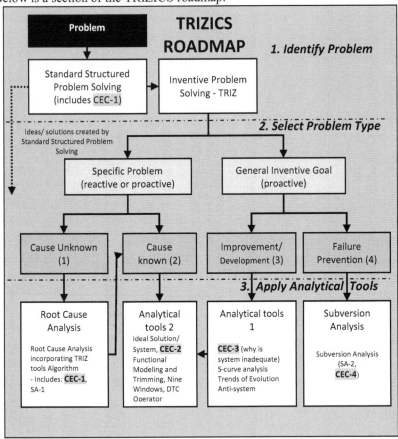

In our TRIZICS Roadmap we define four problem types. We select which analytical tool to use depending on the problem type.

A specific problem may be reactive or proactive and the root cause may be known or unknown. A technical problem where the goal is general improvement is always proactive. General improvement involves improving, developing or inventing a technical system or technical process.

We define preventing failures as a category of general proactive problem solving. If a specific failure mode is targeted for elimination we can use the same tools as if it were a "cause unknown" reactive problem.

Four Types of Technical Problem		
Specific	Cause Unknown	Type 1
(reactive or proactive)	Cause Known	Type 2
General	General Improvement	Type 3
(proactive)	Failure Prevention	Type 4

Specific Problem (may be reactive or proactive)

Type 1: Solve a specific problem when the root cause is unknown.
Type 2: Solve a specific problem for which the cause is known.

General Problem (proactive)

Type 3: Improve, develop, invent a technical system, or technical process.
Type 4: Prevent future failures for a technical system or technical process.

Examples of the four problem types:

Type 1: Eliminate intermittent leaking of a pipe in a gas supply system (specific, cause unknown).
Type 2: Eliminate the fracturing of a glass tube due to thermal expansion when it is heated (specific, cause known).
Type 3: Determine how to develop a motor car to gain market advantage. Invent a better floor cleaner. Improve the efficiency, repairability and quality of a preventative maintenance process (general inventive goal to improve a technical system or technical process).
Type 4: Eliminate the causes of failure for a metal electroplating process, a radio, roller coaster, a kettle (failure prevention).

The way we use Cause-Effect Chain analysis is slightly different for each problem type. We separate CEC analysis into four separate, but very similar tools, one for each of the problem types.

CEC-1 is used to perform controlled brainstorming of the **root cause.**
CEC-2 is used to perform controlled brainstorming to check the **root cause** when root cause is believed to be known.
CEC-3 is used to try to identify problems for **improvement/development** of a technical system or technical process.
CEC-4 is used to anticipate failures; it is also called **subversion analysis.**

The mechanism of how each operates is very similar; the differences are due to the different types of original question that we ask and how we deal with the ends of the chains. Let's start with an example of CEC-1.

1.1 CEC-1: For Finding Root Cause (Type 1 problem)

We recommend that the user applies CEC-1 instead of brainstorming or Fishbone analysis when initially applying Standard Structured Problem Solving (see Chapter 2). Let's use the simple specific (reactive) problem, "my bicycle tire is flat" as an example. To create a Cause-Effect Chain we start with a **target problem**. The problem is clearly stated in the first problem "box."

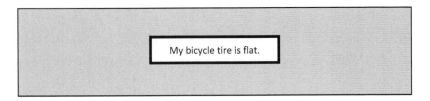

The next step is to "brainstorm" possible reasons why. We simply ask why and each possible answer inserted in a new box and is then connected to the first problem box.

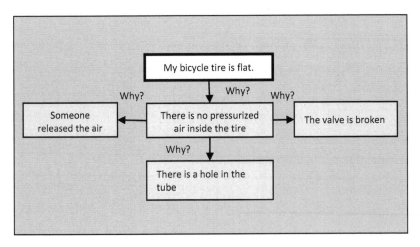

Typically the word "why?" is written next to each connector arrow. This illustrates that the preceding box in the chain is treated as a problem. The "why" is the connector and the box it connects to is the answer, the "because."

More than one connector indicates there is more than one possible answer. These can be individual answers (answer "A" OR answer "B" OR answer "C", etc), or combined answers (answer "A AND B" or "A AND B AND C", etc.).

In the above example, there is no pressurized air inside the tire, we brainstormed three reasons: someone released the air, OR there is a hole in the tube OR the valve is broken. The convention is not to write the word "OR" in the CEC diagram. The "OR" is implied by the three separate connectors that emanate from the box.

To denote an "AND" situation we *do* include the word AND in the box in the CEC diagram. For example, I have a cup of coffee, it cools very quickly because the cup is poorly insulated and the surrounding atmosphere is very cold. Good insulation would have prevented the rapid cooling or a high surrounding temperature would have prevented the rapid cooling.

Both poor insulation AND low surrounding temperature were needed to cause the problem. An AND is indicated on the CEC diagram using an AND box between both of the required causes.

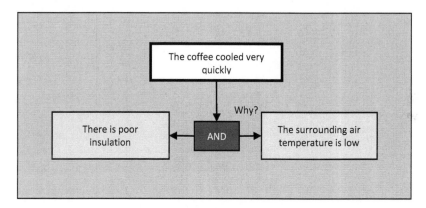

Several boxes may be connected to the AND box if there are several required causes. Eliminating or solving any one is a way to solving the target problem.

Note that sometimes loops can form. For example, a vessel is used to hold a hot liquid. The liquid starts to leak from the seal. Why?

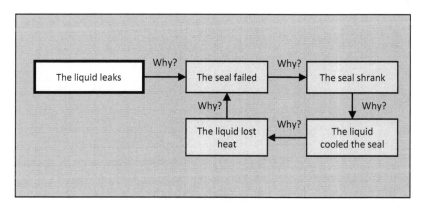

The seal fails because it leaks. It leaks because it shrinks. It shrinks because the hot liquid cools. The liquid cools because it loses heat, and it loses heat because the seal leaks. Breaking the loop by solving any one of the causes solves the problem associated with that CEC.

Let's go back to our flat tire problem. Answers (ideas or reasons) are created for each box. The chain grows and more boxes are added as more causes are identified.

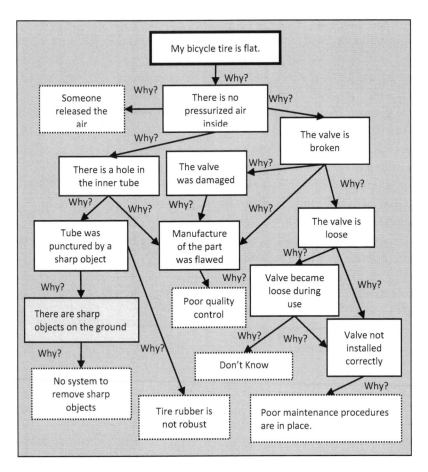

The chain continues to grow until a terminal box is reached. Chain links end for several reasons:

1. There is a physical reason (the tire rubber is not robust, why? Because the rubber material used is not robust, the material has a physical limitation) or other project limitation (for example, tires are not solid, why? It is not legal to use solid tires, there is a legal limitation).
2. There is a procedural reason such as poor maintenance procedures or poor quality control.
3. No more ideas. For example, you may not have any ideas how the valve can become loose. This is a positive result of the CEC tool, it identifies a gap in your knowledge, the next step would be to close that gap and find out the reason.

After completing the root cause analysis using CEC, the next step is to review the chain diagram and create a list of the causes, and then create a plan to address these potential causes.

The plan typically includes actions to perform, such as completing experiments, collecting data to trying to validate the root cause (or causes) or implementing and testing possible solutions.

A routine solution (non breakthrough) may be to simply improve maintenance procedures or quality control, or drive only where there are no sharp objects. As with any analytical tool, the specific problem that is created can subsequently be restated in the form of one or more of the classical TRIZ tool formats. Try to form them into contradictions if possible for more innovative solutions (see Chapter 9). For example, for the answer "tire rubber is not robust," we can create a contradiction: tire rubber should be soft to provide good traction with the ground but hard to avoid punctures.

Each possible cause can be prioritized and pursued until a solution is found that satisfactorily eliminates the problem.

Example of plan: Flat Tire

	Possible Cause (from CEC-1)	Actions	Priority	Completion
1	Someone released the air in the tire.	N/A have validated it was not done on purpose.	N/A	N/A
2	No system to remove sharp objects.	Find out if there is a way of removing or reducing the possibility of sharp objects where I cycle. Are there any scientific effects can help remove them?	D	Sept
3	Tire rubber is not robust.	Solve the contradiction: tire rubber should be soft to provide good traction with the ground but hard to avoid punctures.	A	Jun
4	Poor quality control during the manufacture of the valve and inner tube.	Go to the manufacturer, audit the quality control process and try to develop procedures to meet 100% quality control.	C	August
5	Valve became loose during use (reason unknown).	Create models (scenarios) of how the valve could become loose. Perform experiments and try to loosen them during use.	B	April
6	Poor maintenance procedures in place. The valve was not installed correctly.	Audit and improve maintenance procedures.	A	March

CEC-1 is a tool for logically thinking through to potential root causes rather than random brainstorming. It enables us to create a specific plan. Some common issues that should be avoided when building a CEC-1 are:

1. Do not over-abbreviate. Make sure someone unfamiliar with the terminology and problem could understand and follow the logic of the chain. The chain is also a useful communication tool for presenting your problem solving process to others. If others understand it, they can contribute ideas. For example, don't write "valve" in the box that should say "the valve is loose." Be specific, over-abbreviation leads to inaccuracy and missed opportunities.

2. Do not use the tool to document or justify what you already think. It is a tool for thinking, it should release psychological inertia.

3. Check steps in the chain were not skipped by tracing the logic back from each terminal box. Each box should be the response to the previous one in the chain to which it is linked. If it is not a direct response then a step may have been skipped and potential solutions overlooked.

4. Revise the diagram regularly. Often ideas occur that were not created the first time the diagram was prepared. The CEC is a thinking tool with which you should actively interact.

Try this simple exercise: you walk into a room, turn on the light switch and no light comes on. Why?

Use CEC-1 to identify possible reasons and create a plan for what to do about it. Start with the target problem: the light did not come on.

1.2 CEC-2: To Validate Root Cause (Type 2 problem)

CEC-2 is similar to CEC-1. It does not have to be repeated if it was done as part of root cause analysis for a Type 1 problem.

CEC-2 is performed as an analytical step for a Type 2 problem where root cause is believed to be known. Sometimes root cause is incorrectly identified and we are not working on the right problem or are working on a symptom rather than the cause. For example, if the problem is "I feel cold." I assume that the problem is I am not insulated and so I look for a solution. I try to find a blanket to keep warm. This addresses the symptom. Actually the problem is that the fire went out. I should be working on keeping the fire burning. CEC-2 would have revealed the true root cause.

Before trying to solve a problem by moving to the solutions steps it is important to complete the problem analysis CEC-2 step and not assume the problem is correctly defined.

1.3 CEC-3: Proactive Improvement of a Technical System or Technical Process (Type 3 problem)

As mentioned above, CEC analysis can also be used for general proactive problems. Instead of the target problem being a specific problem, we simply identify the system or process we would like to improve and state the target problem as "the technical system or technical process is not ideal." We use CEC as a tool to identify problems with our system.

For example, let's choose a toothbrush as our technical system. Our initial problem statement is "the toothbrush is not ideal." We identify the gaps to ideality: it does not clean well between teeth, takes too long, etc. So we can build a CEC, define the problems with the system and identify the gaps we would like to close.

Example: a toothbrush is not ideal.

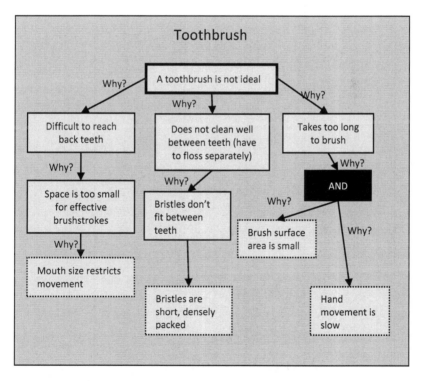

Now create a prioritized list for improving a toothbrush based on the reasons you created. Try to turn them into contradictions if possible.

Forming contradictions is the TRIZ method of creating breakthrough thinking. How to form and solve them will be discussed in Chapter 9.

Example: a toothbrush is not ideal

	Cause	Problem	Contradiction
1	Brush is too big for available space in mouth.	How to make it easy to remove plaque from back teeth.	If brush is large then it cleans back teeth quickly but it can't fit in a person's mouth.
2	Bristles are short, densely packed and thin not long separate and rough.	How to get bristles to brush between teeth.	If bristles are long thin and separated they clean between teeth but they don't clean the surface well.
3	Brush is small and hand movement is slow.	How to make a plaque removal system that will increase cleaning rate.	If speed and size of hand brushing increases then plaque is removed quicker but there is much discomfort.

Exercise: Use CEC-3 analysis to create a list of problems that if solved would improve a flashlight. Start with the target problem statement: a flashlight is not ideal.

Next, try CEC-3 analysis for a vacuum cleaner then for an electric cooling fan. CEC-3 analysis is a key tool for the inventor because we evaluate an existing system and create what problems need to be solved to improve it.

1.4 CEC-4: Prevent Future Failures for a Technical System or Technical Process (Type 4 problem)

CEC-4 is used as part of the procedure to predict future failures, (see Subversion Analysis SA-2 Chapter 10). In this process a list of potential problems is created by the problem statement "the system failed" or "the system is inadequate" in the first box in our chain. Users build a CEC and the possible reasons for failure are revealed. A plan is developed to implement solutions ahead of time to proactively prevent problems. Proactive problem solving is more effective than reactive.

Chapter 4

Ideality, S-curve Analysis and the Trends of Evolution

Chapter Contents

Chapter Summary: *Technical systems are created to provide one or more functions. Systems typically evolve by increasing functionality and decreasing costs, where "costs" means harmful or negative factors in addition to monetary value. It is useful to define "ideality" as the ratio of functionality divided by cost. Systems evolve by following a trend of increasing ideality by increasing functionality and/or reducing cost and harmful factors. The ideal system is one which has infinite functionality and zero cost. The concept of the ideal solution or ideal system can be used as a thinking tool to release psychological inertia.*
The life cycle of a technical system, component or subsystem can be described in terms of an S-shaped curve that develops in four main stages: infancy, growth, maturity and decline. The trend or "law" of increasing ideality drives

a system up along the S-curve. Identifying the position of a product or its parameters on the S-curve can be used to direct development of the product or parameter.

Repeating trends or patterns that describe how systems evolve were identified from the study of patents. Technical systems tend to follow specific paths of evolution rather than develop randomly. We discuss eight trends (plus one) although many more are documented in the TRIZ body of work.

Knowing these trends helps provide breakthrough ideas that can drive a system along the S-curve to a more evolved state. A system, component or subsystem can be evaluated for its own level of development in relation to each "law." This evaluation provides ideas for the direction of development of the system, component or subsystem based on how others have developed in the past. Specific problems may be solved by reviewing the trends for ideas.

1. Ideality

Technical systems, products, tools, etc. are made in order to perform one or more functions. A function is what a system does that is useful. Consumers purchase products for their functionality. We use a hammer to drive in nails, a lawnmower to cut grass, and a computer to provide information. To drive in nails, cut grass and to provide information are all functions provided by technical systems. The function of an MP3 player, CD player, cassette player, all deliver the same main function – to play music.

"Functionality" is the amount of desired useful functions that a system provides. A Swiss Army knife has many different tool attachments and provides more functionality than a simple pen knife with a single blade.

1.1 The Trend of Increasing Ideality states that over time, technical systems develop towards increased ideality. Ideality is functionality minus what has to be paid to obtain the functionality – cost plus negative or harmful effects. Cost is defined not only in monetary cost but, the use of any resources or materials, the number of components used, the amount of space or time used. Harmful effects may be any negative effect associated with the system. For example, a lawnmower cuts grass but it may also be very noisy, which can be considered as a harmful effect. We can think of cost plus harmful effects as the expense we pay for the functionality.

A useful qualitative way of describing ideality is to use the following formula:

$$Ideality \sim \frac{Functionality}{Cost + Harmful\ Effects} \sim \frac{Functionality}{Expense} \sim Value$$

- Ideality (the value of the system) will increase if functionality increases.
- Ideality will increase if the cost and harmful effects (expense) decreases.

To increase functionality, the quantity, quality and diversity of functions may increase. For example, a lawnmower that cuts more grass more evenly provides more functionality. The simple mechanical push lawnmower added an engine to increase the speed of cutting grass. Collectors were added to gather cut grass to increase the diversity of functions.

To decrease cost and harmful effects (expense), reduce the monetary cost, any resources or materials, the number of components, the amount of space or time used and/or eliminate any harmful or undesired effects.

The trend towards ideality drives the development of technical systems. The basic hammer developed into a claw hammer which added the function to remove nails by as well as drive them in. The Swiss Army knife develops by adding a greater number of functions. Computers increased the number of software applications. Originally used for performing data calculations, computers are now used to communicate and provide entertainment while using less space and at less cost.

The trend of increasing ideality is driven by basic economics. We want more for less. Systems with more value in a competitive economic system will survive and prosper.

Now let's consider the notion of the ideal system. The ideal system is one for which functionality is infinite and the cost payment factors are zero.

An ideal system can perform any and all functions free of charge. Of course there is no such thing; there are only imaginary wishing wells and magic wands.

Some systems can, however, have zero cost. For example, if I have a projector and I need a screen to project the image onto, I could buy a screen. But if a wall was available, the image could be projected onto the wall at zero cost. The wall is the ideal screen. The functionality of the screen was obtained for free by using an existing resource. Cost is minimized when existing resources can be used, or there are minimal changes to the system. Using available existing resources is a method TRIZ uses to try to get closer to creating an ideal system or solution.

The concept of ideality itself can be used as a creative thinking tool. As part of our problem solving process we consider **the ideal system** and **the ideal solution.** Brainstorming ideas at an early stage of the problem solving process using the concept of the ideal system and the ideal solution releases significant psychological inertia to provide innovative ideas of solutions or re-direct thinking.

1.2 Using Ideality (the Ideal System) to solve a General Problem (Type 3)

Psychological inertia is released by using the concept of **the ideal system**. The ideal system (or process) performs the function (or functions) by itself for free. The ideal system would have zero cost, no harmful effects and it would have infinite functionality.

1.3 Algorithm for applying the Ideal System to a General Problem
(Type 3)

Step 1 Describe the ideal system (or process).
- Note that the ideal system is no system; the ideal process is no process the functionality is provided for free, no costs, and no resources would be used if the system were ideal.

Step 2 Using the concept of the ideal system/process. Brainstorm ideas for how to improve the ideality of the system/process.

Step 3 Using the list of ideas of the improvements; create a list of solutions to improve the system or process.

For example: I have to improve a preventative maintenance process. Use the concept of ideality (the ideal process) to create ideas for improving the process.

Step 1 Describe the ideal system (or process):
- The ideal preventative maintenance process is no preventative maintenance process.

Step 2 Using the concept of the ideal system/process, brainstorm ideas for how to improve the ideality of the system/process.
Can we eliminate the maintenance? Can we remove specific steps that are not required or combine steps to make it more efficient? Can the maintenance perform itself?

- Eliminate the need to replace specific parts at a high frequency.
- Perform events in parallel.
- Perform more effective inspections prior to completion.

Step 3 Using the list of ideas of the improvements; create a list of solutions to improve the system or process.

- Clean component x every 10 times instead of every time.
- Use a gas cooling liquid to cool down parts quicker.
- Use ultraviolet light to inspect for quality.

1.4 Using Ideality (the Ideal Solution) to Solve a Specific Problem (Type 2)

Similarly we can use the concept ideality to help find solutions to *specific* problems. Significant psychological inertia can be released by using the concept of the **ideal solution**. Just as the best system is the most ideal, the best solution is most ideal. The ideal solution occurs when a system solves a problem by itself, for free. For example, my bicycle tire frequently gets punctures. The solution is to find a way to make a self sealing tire or one that cannot puncture.

1.5 The Ideal Solution Statement

How can the system eliminate the problem by itself, using only the available resources of the system, surrounding environment and free or inexpensive resources?

(Note that we will distinguish between the ideal solution and the Ideal Final Result (the IFR). There are two very specific statement formulations in ARIZ: IFR-1 and IFR-2. They should not be confused with the ideal system/ideal solution statements).

1.6 Algorithm for applying the Ideal Solution to a Specific Problem
(Type 2)

Step 1 State the ideal solution, one where the problem solves itself. How can the system eliminate the problem by itself, using only the available resources of the system, surrounding environment and free or inexpensive resources?

Step 2 Using the concept of the ideal solution, brainstorm ideas for how to solve the problem for free or using the systems local environment or inexpensive resources.

Step 3 Using the list of ideas; create a list of solutions to solve the problem.

Example

Water containing gravel passes through a pipe but the pipe is quickly eroded inside by the gravel. How do we stop the erosion? Use the concept of ideality to obtain solutions.

Step 1 State the ideal solution, one where the problem solves itself. How can the system eliminate the problem by itself, using only the available resources of the system, surrounding environment and free or inexpensive resources?

- The problem is solved by itself using available or inexpensive resources.

Step 2 Using the concept of the ideal solution, brainstorm ideas for how to solve the problem for free or using the systems local environment or inexpensive resources

- The water itself eliminates the erosion.

Step 3 Using the list of ideas; create a list of solutions to solve the problem

- Cool the pipe exterior using a refrigeration mechanism so the water close to the pipe inner surface becomes ice. The ice protects the inner pipe walls.

2. S-curve Analysis

Now that we understand ideality we can describe S-curves. S-curves are a way of describing the life cycle of a technical system, from its initial creation to its eventual decline and replacement by newer technical systems. A typical system has four stages: infancy, growth, maturity and decline. Over time (typically years for most products) the performance of the main functional parameters of a system follows an S-shaped curve.

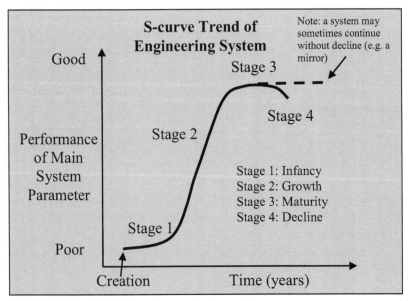

62

"Performance of Main System Parameter" is an assessment of how well a system performs in terms of how effectively it delivers its main function. For example, if the system is a kettle, the main function is to heat water. The main parameter delivered, is the heating rate of water. Other parameters the system provides are the quantity of water it can heat, the heat loss rate after boiling and the insulation capability of the handle. We assess these other parameters of the system in terms of their own performance separately.

The performance of each parameter may be at a different stage of development and hence at a different stage of the S-curve.

Individual components, subsystems, etc. can have their own S-curve and be at a different stage on the S-curve. For example, for a motor car the stage of an internal combustion engines position on the S-curve may be different to its tires, dashboard display or seats. Each system is at a different stage of its developmental life cycle.

A good way to think of the y-axis is "level of ideality." How ideal the system is in terms of functionality divided by cost.

At the start of the life cycle, ideality is low because a new system typically has low functionality and high costs. Research costs are high; the system still has many technical functionality issues and may have unresolved harmful effects. As it progresses up the S-curve, functionality improves and the technical problems are solved, incremental technical improvements are made. At maturity ideality levels off, the system is not developed further, functionality, costs and harmful effects are stable. In Stage 4 the system declines, new alternative competing systems with new technologies that are more ideal and provide greater value, force the mature system out, for example MP3's replaced CD's.

At each stage business strategies can be developed in terms of ideality. At Stage 1, focus on technical problems to provide functionality. During Stage 2, work on optimization making incremental improvements and some cost improvements. In Stage 3, focus on reducing costs, functionality can't be significantly increased, try to develop a new technology, or solve contradictions to create a breakthrough so you can move to a new S-curve. Stage 4, reduce costs, try to find specialized markets, and work on contradictions to create a new technology or product.

Before further discussing the S-curve, let's add three more curves below the S-curve. It is useful to plot the Number of Inventions, the Level of Invention and the Profit levels for each of the four stages of the S-curve. Typically these parameters change in characteristic patterns as the life-cycle of a system, component, and subsystem evolves.

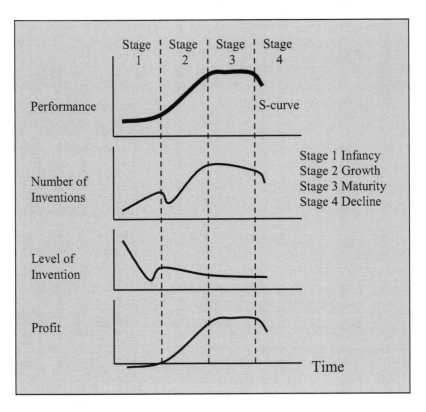

2.1 S-curve Stages

Stage 1: Infancy

Creation is when breakthrough invention leads to the appearance of a new system. Typically a high level solution solves a contradiction which delivers new functionality or a new or potentially more ideal way of delivering an existing function. A Level 5 solution may lead to a new industry for example the solid state transistor led to the semiconductor industry and manufacture of integrated circuits.

After the system is born, from infancy through all later stages, the system evolves. The trend of increasing ideality drives system development and hence up along the S-curve.

$$Ideality \sim \frac{Functionality}{Expense}$$

During Stage 1, the system is new and often exists only as a prototype. Expenses (cost and harmful effects) are high. The system has many unresolved technical problems is unreliable and inefficient. Parts and design need further refinement. Safety and environmental or non-technical requirements such as legal issues may stop the product going to market. The system lacks resources (people and cash) for development projects which causes slow growth. Costs are usually incurred with no profit earned.

As Stage 1 progresses the number of inventions generally increases to improve the efficiency and design of the system. These are initially higher level breakthrough solutions to make the system more functional then the inventions become lower level as more incremental improvements are made. The Level of Invention often increases to meet the functionality needed to bring the product to market.

Stage 2: Growth

By the end of Stage 1, many of the technical issues are now resolved. Cost and harmful effects have been reduced. As it proceeds through Stage 2, the system gains greater market acceptance. Often there are new areas of application. The system becomes profitable and continues to develop as more resources become available. Rapid growth occurs, increasing production volume and profit. The number of inventions grows. Typically these are low level incremental improvements – Level 1 or 2.

Note that if other suppliers are present with similar competing systems at the transition to Stage 2, it is important to be first to market and maintain market dominance otherwise the other competitors will quickly reduce profits as they enter the market.

Stage 3: Maturity

Growth limits have been reached. The system development is flat or there is very slow growth because the system has reached the technical or engineering limit of development. Production volume stabilizes. Profits remain high and stable. The level of inventions continues to be low, and quantity of inventions high (low level solutions already known within the industry are documented).

Causes of Stage 3 Leveling
The system flattens out because it has reached a development limit or costs become too great to justify improvements. Developmental limits include the physical limits of technology or engineering development, the infrastructure and environment (supersystem) and legal or other non-technical limitations. Low level improvements are made; to grow would require higher level solutions to more difficult contradictions.

For example, the speed of communication is limited by speed of light. The rate of heating of water by an electric kettle is limited by the specific heat capacity of water. The capacity of a bus is limited by the supersystem, the size of bridges the width of roads. The size of combustion engines is limited by the amount of pollution, mileage efficiency, and price of gas. The speed of a train is limited by safety concerns. Only low level improvements are made to these systems, to grow would require new technology and moving to a new S-curve.

Stage 4: Decline

Some products might stay in Stage 3 indefinitely. They have been around for a long time with no significant improvement or decline. But typically systems decline.

Causes of Stage 4 decline

Production volume falls, driven by falling demand usually because of competition from a new system reaching Stage 2 (compact discs replaced vinyl records) or there is some change in the infrastructure or environment (supersystem) that drives a change in demand, (the introduction of digital cameras eliminated need for photographic film or a ban on sale of tobacco reduced the need for lighters).

System Revival

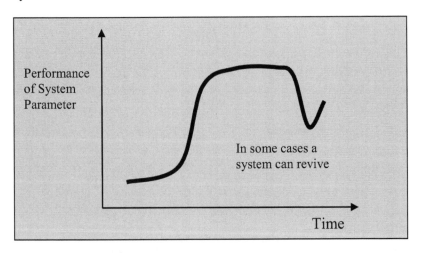

A new situation can emerge where the old system can become useful again or a new technology or material is found. Vinyl records are now popular with DJ's, body armor was abandoned for being too heavy for battle but returned with the invention of light strong composite materials and is used by riot police.

Evolution to new S-curves

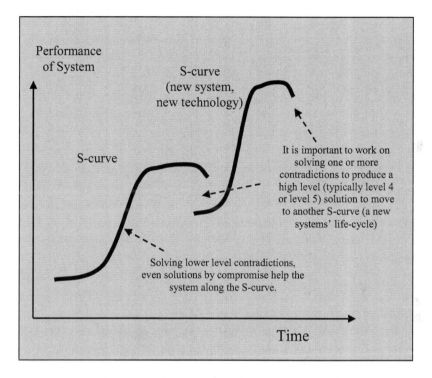

At late Stage 3 or in Stage 4, it is important to work on creating a new system. This requires solving one (or more) contradictions to produce a high level solution, typically a Level 4 or Level 5 solution. Often the new product initially has less functionality than the old one but has a lot of potential and the fall in system performance is acceptable to achieve future gain (system is less ideal, it has less functionality, higher cost and harmful effects, it provides less value to the customer, usually profits are small or there is a loss because the system has returned to Stage 1 of the S-curve).

Of course the S-curve will not have exactly the same proportions from one curve to the next. Times, rates of development and decline will vary and the general shape may change. There may be multiple S-curves over the long term, not just two. The important point is during Stage 3 or 4; a new high level solution that solves a contradiction creates a new, improved method for achieving the same function of the initial system.

Contained in the TRIZ body of work there are documented lists of typical indicators used to identify at which stage a technical system is in, although most manufactures and developers are aware of the stage of their system. Each stage has a corresponding list of recommended actions for how to develop the system along the S-curve when it is in a specific stage.

For example, a typical indicator for Stage 3 is that further development may be limited by the supersystem. Microwave ovens are in Stage 3; their power cannot be substantially increased because they would potentially cause explosions. To improve ideality, improve aesthetic design is recommended, (thus reducing what may be considered as the harmful effect of how the ovens look), reduce costs or add more diverse functionality (add the ability to grill) and to try work on a breakthrough for a new technology (a new S-curve).

3. Trends of Evolution of Technical Systems

System advancement is driven by the trend of increasing ideality. Increased value is the basic reason *why* systems evolve. The drive towards greater value is the force that drives a technical system up the S-curve.

The study of hundreds of thousands of patents revealed a number of repeating trends that indicate *how* technical systems evolve. The evolution of technical systems is not random but is governed by objective laws. These trends show how technical systems have developed in the past. They can therefore be used to forecast how systems will develop in the future.

During their evolution, systems progress both by making basic incremental improvements (implementing compromise or Level 1 type solutions) and by overcoming contradictions (Level 2 and higher solutions) to make more significant breakthrough solutions. Knowing the trends can help provide a solution that advances a system along the S-curve or leaps from one S-curve to a new one.

Trends of evolution are mainly used for general non-specific (Type 3) proactive problems where we are trying to improve a system. The trends show the most likely path for technical development of a product. They allow us to forecast the future technical changes in a system that can be exploited to maintain market leadership, obtain early patents, etc. The trends can also be used to create ideas to provide solutions to specific technical problems (Type 2).

We will discuss how to apply the trends as an analytical and problem solving tool after we review each trend in detail.

The trends are often referred to as "Laws of Evolution." Some are more like laws, others like trends. The first three listed below are like laws. The others are more like trends, and unlike scientific laws, they can be violated. For example, the law of increasing Su-field development states mechanical fields will be replaced by electrical fields. In the case of microelectromechanical systems (MEMs) we could say the silicon chip moved from being an electronic system to mechanical to create these devices.

We will discuss the eight original trends compiled by Altshuller and his colleagues in the 1970's plus the trend of dynamization that was added to the list in the 1980's.

Convention is to class the trend of dynamization as 6a rather than 9 because it is considered to be a trend of Kinematics that occurs during the growth phase (see table below) and those trends were already numbered 4, 5 and 6, so adding another to the group made it 6a. To date, many more trends and sub-trends (trends within trends) have been recognized, and more continue to be added to the TRIZ body of work. We consider these "eight plus one" only.

Altshuller classified the original eight trends in terms of a products life cycle. The three groups of trends were named Statics (trends 1-3), Kinematics (trends 4-6a) and Dynamics (trends 7-8).

Stage	Trend
Creation (Statics)	1,2,3
Growth (Kinematics)	4,5,6,6a
(Dynamics)	7,8

Statics (1-3)

Statics are trends that relate to the creation of a technical system.

Kinematics (4-6a)

Kinematics are trends that govern the development and maturing of any technical system.

Dynamics (7-8)

Dynamics are contemporary trends related to technical systems under specific physical circumstances.

Statics and Kinematics apply to the development of any system in general, not only to technical systems, for example biological systems. Altshuller indicated Dynamics are contemporary trends related to technical systems under specific physical circumstances.

3.1 Details of the Trends of Evolution of a Technical System

Altshuller's Eight Trends (plus one)

Statics (1-3)
Statics trends relate to the creation of a technical system.

69

1. The Trend of Increasing Completeness of a System

A system requires four parts or "functional blocks" to be viable, the Operating Part, Transmission, Energy Source and Control System.

1. Operating Part: the part (or parts) that perform the *main function*. For an electric drill, the drill bit is the operating part.

2. Transmission: a means for converting energy such that it is transferred to the operating part.

3. An Energy Source: a source of energy to drive the system, for example an electric motor.

4. A Control System: a means of directing at least one system part to perform its function.

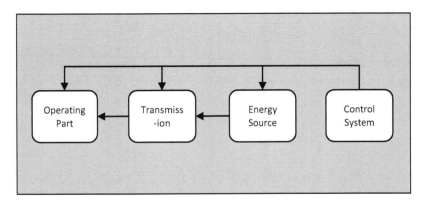

For example, a chair has an operating part (the seat) a transmission (transfer of weight from person to seat), energy source (gravity) and control system (person's brain). A modern sewing machine has an operating part (needle), transmission (gears for moving the needle), energy source (electric motor), control system (a person or onboard computer to direct actions).

All technical systems must possess these four functional blocks in order to be a viable technical system and at least one of its functional blocks must be controllable. An engine is of no use on its own, but when installed in a motor car it provides energy to drive the car. The car also needs a drive shaft to transmit the energy into motion; it needs controls, a steering wheel, accelerator, etc. to control it and wheels (operating part) to move it. A hull acquires a rudder and engine and propeller to make a motor boat. This law applies during the creation of a technical system, at Stage 1.

2. The Trend of Energy Conductivity

In order for a system to operate, it needs to provide at least minimal energy that flows through all parts of the system. A system will not operate unless energy flows through all the required parts flow. Here energy includes the flow of information, substances, materials, objects, etc. A system cannot be controlled unless there is energy flow between at least one functional block and the control.

A light bulb won't operate without the unhindered flow of electrical energy. A digital watch will not function unless it is driven by the flow of electrical power. A motor car requires energy to flow from the engine through the wheels in order to move. A freeway system won't operate when there is a blockage on the road. This law is useful during Stage 1 of the S-curve, system creation.

3. The Trend of Increasing Co-ordination (also known as the Trend of Harmonization or Rhythm Coordination)

In order to be viable, the parts of a system must work together with other components of the system, supersystem and their parameters. The system must work in a coordinated manner. Over time systems continue to develop coordination.

- Rhythm co-ordination: the size and shape of a wind powered generator propeller is coordinated with the strength and weight of materials and with the typical wind speed to maximize power generation. The vibration frequency of a jackhammer is coordinated with the removal rate of concrete. An irrigation water supply is turned on at night to minimize evaporation during daytime heat.

- Shapes of objects or parts are coordinated to be compatible with each other and the supersystem. The design of a car seat is made more fitting for the comfort of the human body. Manufacturers develop special boxes for transporting eggs and special tool kits are created for fixing eyeglasses. The computer mouse has become more ergonomic, better fitting a human hand.

- Materials are coordinated; materials with similar coefficient of thermal expansion prevent the separation of layers in semiconductor devices.

Coordination is a necessary consideration at the creation (Stage 1) of a product, however it is also useful during later stages. For example chairs are mature products that are only now becoming more ergonomic.

Statics requires, a technical system must have all four elements, energy in some form must flow through each functioning part, and the parts must work together.

Kinematics (4-6a)
These trends apply to the growth and maturing of a technical system.

4. The Trend of Ideality has been discussed above.

Development of technical systems is driven by the increase in functionality and the reduction of cost and harmful factors. The ideal system is one which provides functionality at zero cost.

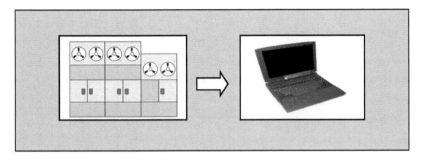

The functionality of a computer developed greater processing power and versatility of applications at reduced cost and physical size.

5. The Trend of Uneven Development of System Components

The development of a system's parts typically grows unevenly – the more complex the system, the more the number of non-uniform developing parts. The uneven development often leads to the creation and opportunity of identifying and solving new contradictions.

- Components may be at different stages of evolution. A sailing ship could move but their navigation systems did not develop to a high level until later. Many explorers did not know where they were. A ship's hull has developed to be large but the braking system has not evolved to effectively stop the ship. It takes several miles to stop a large oil tanker. These discrepancies can be formed into contradictions to be solved.

- There was rapid development in the reduction in size of hand-held electronic devices, but slow development in the means of inputting information. The keyboard was less evolved because the physical contradiction that the keyboard must be large to allow easy input of information and small to be portable was still being solved.

6. The Trend of Transition to the Supersystem

Over time, technical systems merge with the supersystem. As the system reaches its own limits of development in order to become more ideal, it merges with external systems.

The most basic integration with the supersystem is the trend of mono-bi-poly. A single system (mono) integrates with another identical system forming a bi-system to increase its functionality. An example is a single razor adds another to become a twin blade razor. Multiples of the same system can be integrated to increase functionality even further to form a poly system, for example a multi blade razor.

As systems follow the trend of transition to the supersystem, they start to include more diverse types of system. First, they incorporate similar systems that perform a similar function. For example, a knife with different types of blade, then differences begin to grow to include more divergent functionality. For example, a screwdriver for both flat and cross head screws.

Competing systems integrate. For example, the hybrid car has an electric fuel cell and a gas combustion engine. Systems that complement each other in the same process combine. For example, washing powder and scent fragrance. Systems that perform opposite functions combine. For example a pencil with an eraser or a hammer and nail extractor.

The number of systems that integrate increase – a Swiss army knife incorporates blades, bottle opener, nail files, screwdriver, comb, laser pointer, even a flash memory device etc. Printers have been integrated with photocopiers, fax machines, scanners and many even have a stapler and staple extractor.

Distinctions between the system's main function and the function of the systems integrated with the supersystem often become wider. For example, an integrated bottle opener and radio.

6a. The Trend of Dynamization

As a system develops it will trend toward an increase in the ability to change one or more of its parameters in time. Trend 6a was added after the initial eight trends. Since it belongs to the Kinematics group, which was initially 4, 5 and 6, it is known as 6a.

Solid systems and materials become more flexible; they will become more dynamic and gain more freedom of movement. Solid systems of components evolve to include a single joint, then multiple joints. For example, a ruler was originally solid then evolved to have one joint then many joints to become a folding ruler. Then it became flexible tape measure, and eventually used a field (laser measure).

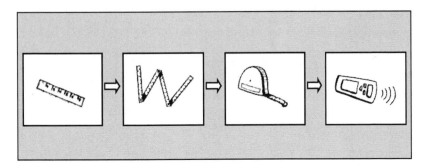

- Systems may develop from being in a solid state to be in the form of powder, then liquid, then gas and finally as a type of field.

- Fields typically evolve from constant to gradient, impulse, resonance, interference.

- Freedom of movement develops from 1 dimension to 2 and then to 3 dimensions. Communication by telephone cables (2D) moved to 3D wireless.

Dynamics (7-8)

Altshuller indicated Dynamics are contemporary trends related to technical systems under specific physical circumstances.

7. The Trend of Transition to Micro Level

The operating part of a system begins as a macro level component or subsystem and develops towards the micro level. The operating part is the bristles on a paintbrush, the propeller of an airplane, the blade of a saw. Instead of macro objects, the work carried out by the operating part is performed by a new part or subsystem that operates at the micro level, the level of particles, atoms, molecules, electrons or particle fields. The transition from single electric valve transistors to semiconductor chips containing hundreds of millions of transistors is a prime example of transition to the micro level.

A beam of high pressure water is used to cut metal instead of saw blade, an aerosol spray is used to distribute paint evenly instead of a brush, a jet engine is used instead of a propeller, chemical reactions remove stains instead of physical abrasion, the operation of computers uses electrons semiconductor materials instead of abacuses.

74

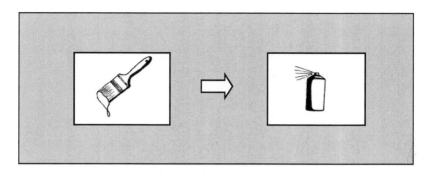

8. The Trend of Increasing Su-field Development

The evolution of technical systems will proceed in the direction of increasing Su-field development.

Su-field is an abbreviated term for substance-field interaction (Su-fields will be discussed in detail in Chapter 8). Typically this interaction is depicted as a triangle in which substance 2 (S2) acts on substance 1 (S1) via a field. S2 and S1 can each be any "thing," such as an entire system, an object, or material, etc. but S2 and S1 cannot be a field. This complete three component interaction (Su-field) is also known as the minimum working "system."

The trend of increasing Su-field development is that incomplete, ineffective or harmful Su-fields strive to become effective complete Su-fields and that effective complete Su-fields will develop by increasing the dispersion of substances (the number and types of "things" will grow), the number of links between the elements will grow (more interactions by forming complex, double and chain Su-fields) and the responsiveness of the system will grow (the ability of the system to be controlled). Systems will also trend to more evolved field types. Mechanical fields evolve to acoustic to thermal to chemical and finally to electric and magnetic fields (sometimes the mnemonic MATCHEM is used to help remember this sequence).

How Su-fields are developed to be made effective, become more evolved and ideal is described in the list of 76 inventive standard solutions (see Chapter 8). There are many specific trends that drive Su-field development discussed in the 76 inventive standards that are discussed in the trends above. For example, transition to the supersystem, transition to micro level, dynamization, trend to more controlled fields, etc.

| Insufficient Su-field | Chain Su-field |

Let's take a specific example for how systems evolve via the trend of increasing Su-field development. Fluid (S2) used in a car suspension system insufficiently dampens mechanical movement (F1) in a car (S1); it is not effective over different types of terrain. The system evolves by changing the fluid inside the piston such that it contains ferromagnetic particles. The modified fluid (S2') can change the fluid's viscosity via an electromagnet (S3) which applies a variable magnetic field (F2) increasing the control of dampening by the suspension for different types of terrain. The system evolved by increasing the number of elements (substances) and fields.

3.2 Algorithm for Applying Trends of Evolution to a General Problem

1. Describe the system or component that is to be developed.
   Note that a system may have several components each at a different level of development relative to each trend.
2. On an arbitrary scale of 1-10 (1 is low 10 is highly developed), estimate the level of development relative to each trend.
3. Based on the level of development, define a plan for development of the system. The plan should focus on improving the less developed aspects of the system by bringing them to a more evolved state.
4. Repeat for each component of the system.

	Trend of Evolution	Level	Plans to Develop Trend
Statics			
1	System Completeness		
2	Energy Conductivity (optimization of flows)		
3	Coordination		
Kinematics			
4	The other trends are the mechanism for 4. Law of Ideality.	N/A	N/A
5	Evenness of development		
6	Transition to Supersystem		
6a	Dynamization		
Dynamics			
7	Transition to Micro Level		
8	Increasing Su-field development		

- It may be useful for the user to include other trends in the table by reviewing the TRIZ body of work.

76

3.3 Algorithm for Applying Trends of Evolution to a Specific Problem

1. Describe the specific problem.
2. Review each trend and use it to try to prompt an idea of the solution

For Example: Conductive pins must contact metal pads on a semiconductor chip for testing the electrical performance of the circuitry. The thickness of the metal is only a few nanometers and the contact must be very accurate. If the pins penetrate too deeply they cause damage to the pads on the chip.

Solution: We can follow the trend of dynamization (6a) and "dynamize" the pins. Attach them to a flexible material that limits the force applied by the pins to the pad so that the pins will contact the surface but at a very low, non-damaging level.

Chapter 5

Nine Windows the Anti-system and the DTC Operator

Chapter Contents

1. **Introduction**
2. **Nine Windows**
   2.1. Creating and Using a Nine Windows Diagram
      o General problem (Type 3)
      o Specific problem (Type 2)
   2.2. Using Nine Windows as a Planning Tool
   2.3. Algorithm for Nine Windows
3. **The Anti-system**
   3.1. The Anti-system Algorithm (single system)
   3.2. The Anti-system Algorithm (Nine Windows)
4. **The DTC Operator**
   4.1. The DTC Operator Algorithm

Chapter Summary:

Nine Windows *is also known as the System Operator or Nine Screens. Nine Windows helps us to see a problem in the context of the past and future, the environment in which it functions (supersystem), and its parts (subsystem). This prompts us to consider solutions in the past and future and at the supersystem and subsystem levels instead of simply focusing the present system level*

To use Nine Windows, write the problem or system name in the center box of a 3-by-3 matrix of nine boxes. The bottom row is subsystem level, middle is the system level, and the top row is the supersystem level. The first column is the past, second column is the present and the third column is the future. When we have identified the supersystem and subsystem components for the past, present and future by entering information into the Nine Windows, we can consider using these as resources to solve our problem, or develop our system. We ask:

- *Can we do something at the subsystem, system or supersystem level **in advance** to fix the problem or improve the system?*
- *Can we do something at the subsystem, system or supersystem level **in the future** to fix the problem or improve the system?*
- *Can we do something at the subsystem, system or supersystem level **in the present** to fix the problem or improve the system?*

Sometimes this is called "thinking in time and space." Of course there are many different levels of system and different scales of time that can be chosen. There may be a sub-subsystem or super-supersystem and so on. Time between past and future could be short - part of a second, or medium or long term – years or decades, or the user may not use a specific scale. We can therefore create many sets of Nine Windows for a particular problem or system.

Prompted by the wider view of the system, the user looks for new ways of solving the problem, developing or planning for improvements to the system.

***The anti-system** is also used to prompt ideas. Psychological inertia is released by considering the anti-function or anti-action of a system, component, part, etc. By identifying the opposite function or action that a system, part, etc., performs and by asking how would such an anti-system work and what could it be used for? We are prompted to create revolutionary new ideas by this radical proposal.*

We can apply the idea of "anti-system" to Nine Windows. For each component in each box, we can consider its anti-function or anti-action then create ideas based on such radical thinking.

***The Dimension Time Cost Operator,** similar to Nine Windows, the DTC Operator (Dimension Time Cost Operator) is a tool for releasing psychological inertia to help with creative thinking. It helps prompt ideas by considering how the problem can be solved if its dimensions, time (including speed), or cost were exaggerated to be extremely large or small.*

1. Introduction

Nine Windows, the **anti-system** and the **Dimension Time Cost** Operator are commonly used tools for releasing the creative imagination when trying to solve a problem.

2. Nine Windows

Nine Windows is a versatile tool used to release psychological inertia. Instead of thinking about a problem (Type 2 specific or a Type 3 non specific inventive goal to improve a system) in the present and at the system level, we consider the system in the past and in the future, and at its subsystem and supersystem levels. When we see the system in the wider context of time and different system levels, we can expand our thinking and see ideas that might not occur to us if we focused only on the problem at the present system level.

Technical systems contain subsystem components and are part of a supersystem (external components the system interacts with).

Supersystem: the external environment and components a system interacts (or can interact) with.
System: a system that was created to perform one or more functions.
Subsystem: a component or parts of a system.

Example: If the technical system is a light-bulb, what are the subsystem and supersystem components?

System: light-bulb
Subsystem: glass, inert gas, filament, screw base.
Supersystem (includes the product): room, air, switch, person (product), etc.

We typically consider and focus on the immediate system to be developed or problem to be solved instead of considering the past, future and system levels. We overlook what is to the sides, above or below. By creating a Nine Windows diagram, we see the situation from a broader perspective; we can see more ways of solving a problem or developing the system as a whole.

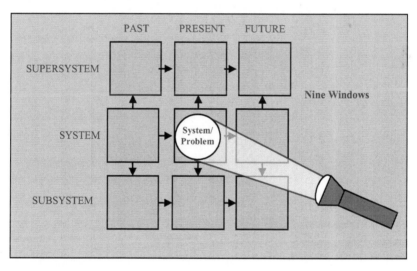

For example, consider a tree. We might think only of a tree with a trunk and branches, maybe we even consider the roots. If we expand our image of a tree, we would consider the tree in the past. We can imagine it as a sapling or even as a seed. If we consider it in the future, it may be timber or a chair, etc. At the macro level the tree may be a copse, a wood or forest. At the micro level it is comprised of leaves, branches, fruit, etc. Let's create a Nine Windows diagram for a tree.

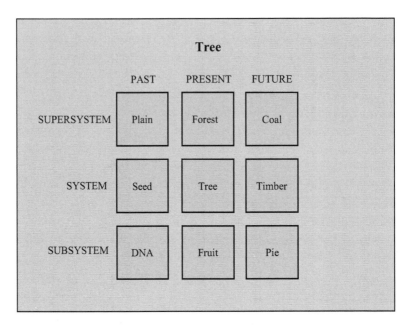

Tree

	PAST	PRESENT	FUTURE
SUPERSYSTEM	Plain	Forest	Coal
SYSTEM	Seed	Tree	Timber
SUBSYSTEM	DNA	Fruit	Pie

We now have a wider conceptual picture of the tree as a system within time and space. In the future a tree will become timber, at the supersystem level, coal. The Nine Windows diagram shows us links. In order to enjoy the fruit pie, we need forests. To have timber, we must plant seeds. We see the tree in the context of its supersystem, subsystem, past and future.

When trying to invent or solve a specific problem, it is useful to create a Nine Windows diagram so the problem solver can see these wider aspects of the system. Potential solutions might lie in the subsystem or supersystem to prevent the problem, or it may be possible to recover or repair the problem after it has occurred.

When developing a system, drawing a Nine Windows diagram may identify changes in the subsystem or supersystem might be needed. For example a triple-decker bus may solve a capacity problem but the bridges (the supersystem) would be too low to allow them to pass underneath.

Altshuller believed that "talented thinkers" thought in the same way as depicted by the Nine Windows diagram, seeing the problem or system to be improved in terms of the past, the future, and at the system, subsystem and supersystem levels. He also believed that talented thinkers use the concept of the "anti-system" to release creative ideas; we'll discuss the anti-system below.

2.1 Creating and Using a Nine Windows Diagram

To create a Nine Windows diagram, start with the center box. In the center box, write the name of the system to be developed (e.g. car, if the system to be developed is a car) or the system where the problem resides and the problem (e.g. a pipe/leaking pipe, if the problem is a leaking pipe). Next, complete the center column. List the sub and supersystem components. Complete the first and third columns by identifying the state of each item in the center column in the past and what they will become in the future.

Note that the user could identify not only the subsystem and supersystem components but there could be sub-subsystem components, super-super system components, etc. Many system levels are possible. The user should decide what is appropriate. We recommend that the user make multiple Nine Windows diagrams if there is a need for additional levels beyond the subsystem and supersystem levels. The user should not try to increase the number of windows to 4x4, 4x3, 5x5, etc. Nine Windows diagrams should use only Nine Windows.

Similarly the user may choose multiple timescales. Typically for a specific problem, a short term timescale is recommended. For general inventive goals it is to choose a long term or a non-quantified timescale to identify the historical developmental state of the system or components in the past. Sometimes it is useful to create multiple diagrams: very short term, medium, very long term, sub-sub, super-super system, etc. The scales should be chosen based upon the goal of the problem.

General Problem

Let's try the example of a car. To start with we will consider the historical states of the car with no fixed timescale then short and long term.

	Past	Present	Future
Supersystem	person, dirt track, fuel depot, weather, small amount of horses and carriages.	Person , road, gasoline station, weather, traffic	person, gas stations provide non-emission fuel or no gas stations, possible climate change, traffic
System	carriage	car	no-driver car non-emission car low fuel consumption multiple passengers
Subsystem	horse, livery	engine, tires, gas, body, seat, steering wheel	non-emission engine, non-wear tires, no gas, low air resistance body design, ergonomic seating,

It is useful to look at the trends of technical system evolution. For example, the trend of decreasing human involvement may lead to no driver cars.

Car: Timescale, short term (+/- 5 minutes)

	Past	Present	Future
Supersystem	person walking towards car, road unchanged, gas station unchanged, traffic without my car	person , road, gasoline station traffic with my car	walking from car, road worn, gas station closed at night, traffic quiet at night
System	parked car	car	parked car
Subsystem	cold engine, cold tires, cold seat, cold steering wheel	engine, tires, gas, body, seat, steering wheel	warm engine, worn tires, pollution, used gas, warm seat, moved steering wheel

The two diagrams change depending on whether you are considering the historical state or fixed timescale of the system. Often it is valuable to create a medium or even long term set of windows.

Car: Timescale, long term (+/- 20years)

	Past	Present	Future
Supersystem	child, dirt track, empty lot	person, road, gasoline station traffic	old person, worn road, abandoned gas station, restricted traffic
System	raw materials	car	scrap or recycled materials
Subsystem	raw materials	engine, tires, gas, body, seat, steering wheel	scrap or recycled materials

The type of solutions and ideas for development may change depending on the timescale and number of system levels evaluated.

With this information, we can predict future developments of the system and supersystem and plan for the changes that are needed ahead of time. Longer term timescales can be useful for proactively identifying and solving contradictions.

Specific Problem

For specific problems, it is possible a solution could be identified in the supersystem level in the past (for example change the humidity in a room to stop corrosion of parts), or at subsystem level in the present, etc. Let's try a specific problem example.

Problem: a car radiator hose pipe is leaking water (the water overheated, and so expanded and caused the leak)
System: hosepipe
Timescale (short term)

	Past	Present	Future
Supersystem	cold (non-operating) radiator, cool engine, cool water, person, tools, air	car radiator, engine, hot water, less water, person, tools, air	car radiator cold, engine cold, water cold, person, tools worn, air
System	pipe close to leaking	**hosepipe**: radiator hosepipe is leaking hot water	pipe repaired with a patch or replacement pipe
Subsystem	hot water, hot pipe, hot paint	water, pipe material, paint on pipe	used water, bonded pipe material

Now try to create solutions for each box, use the resources available if possible:

- Can we do something at the subsystem, system or supersystem level **in advance** to fix the problem?

Example solutions: At the subsystem level we could have used a type of paint that expands and seals leaks when in contact with water or use a liquid that expands less than water or use a thicker pipe. At the system level, we could install a detector that informs us that a pipe is close to leaking so we could replace it before it leaks. At the supersystem level we could reduce the temperature of the hot water by increasing the cooling to the radiator by driving slower.

- Can we do something at the subsystem, system or supersystem level **in the present** to fix the problem?

Example solutions: At the system level, create a twist in the pipe to stop the leak by tension. At the supersystem level, use some of the persons clothing or nearby material to wrap around the pipe and stop the leak, at the subsystem level, can the paint be scraped off and used to plug the leak?

- Can we do something at the subsystem, system or supersystem level **in future** to fix the problem?

Example solutions: At the system level we can repair the pipe with the patch and bonding material or use a replacement pipe, they were not available before the leak, but purchased after. This gives us the idea of making them available as part of the supersystem in advance of future leaks. Using Nine Windows has prompted ideas for planning for future needs and events.

85

2.2 Using Nine Windows as a Planning Tool

Sometimes Nine Windows is used as planning tool. The item that is being planned for is written in the center box. The super and subsystem relating to the item entered into the center upper and lower boxes. The past and future developments or planning goals are entered. The windows are reviewed at various timescales to try to predict future issues, goals and plans are to ensure that system, subsystem and supersystem developments will support the future goals.

For example: plan for an increase in productivity of soup.

	Past	Present	Future
Supersystem	small farms, no factory, local store, plates, bread	farms, factory, transportation, supermarkets, buyers, customers, plates, bread	bigger factory, larger trucks, more supermarket space, more customers, more bread. Bigger market
System	homemade soup	factory made soup	soup is heated and consumed. Eaten in more places. Happy customers
Subsystem	seeds, raw material	vegetables, stock, can	Waste peel, scrap

Above we used a mixed timescale to illustrate the planning capability. If soup production is to increase we should consider how we will provide more seed and raw packaging materials, how to create larger farms, how to deal with more waste, how to have it heated and consumed in more places, how to address the need for more transportation, bigger trucks, competing supermarket space.

Increased soup consumption will lead to more bread consumption, should we manufacture bread in anticipation of increased demand? Instead of focusing on how to make more soup (the here and now) we see the system as a whole by using Nine Windows and can plan appropriately for changes in the sub and supersystem.

2.3 Algorithm for Nine Windows

Step 1 Write the "System" or "System Problem" naming the problem and the system in which the problem resides in the center square.
Step 2 Draw the Nine Windows. Choose a timescale, line of development, or one or more timescales (short, medium, long term).
Step 3 Complete each of the windows.
Step 4 Now try to create solutions:
- Can we do something at the subsystem, system or supersystem level **in advance** to fix the problem?

86

- Can we do something at the subsystem, system or supersystem level **in the present** to fix the problem?
- Can we do something at the subsystem, system or supersystem level **in future** to fix the problem?

Step 5 Document ideas

3. The Anti-system

The anti-system is a tool for prompting radical ideas mainly for Type 3 problems. We identify the anti-function or anti-action provided by objects. For a given system, component, part, etc., we identify its useful function or action, which allows us to state its opposite (anti) function or action. Then we create ideas prompted by the concept of the anti-system which is a system that provides the opposite function or action to the given system. We try to create ideas by asking – how would such an anti-system work, what could it be used for? Often, prompted by these radical notions, we create breakthrough concepts.

Note that we will use the term the "main useful function." The main useful function is the primary function a system, part, component, etc. is used for – its main purpose. See Chapter 6 Section 2.2 for a more detailed definition.

3.1 The Anti-system Algorithm (single system)

1. Name the system, component, part, etc.
2. What is the main useful function of the system, component, part, etc?
3. What is the anti-function (or anti-action)?
4. Describe the anti-system
5. Create ideas, how would such an anti-system work, what could it be used for?

Example 1
1. Sunglasses.
2. To stop (reduce) light.
3. To provide (increase) light.
4. Glasses that provide or enhance light.
5. Night vision goggles, glasses that have LED's that light up when the user is tired, glasses that provide light for reading, glasses that light up to be more easily found.

Example 2
1. Pen.
2. To direct (flow) ink.
3. To remove ink.
4. A pen that removes ink.

5. A pen that removes the white dye from paper. A pen that removes dye by heating (burning) would never need ink.

Some users apply the concept of the anti-system to Nine Windows to try to prompt ideas.

3.2 The Anti-system Algorithm (Nine Windows)

First a standard Nine Windows analysis diagram is created. Then a second anti-system Nine Windows diagram is created in which each object in each window provides the opposite function to the original.

1. Create a Nine Windows diagram

2. Create a second anti-system Nine Windows diagram is created in which each object in each window provides the opposite **function, action or state** to the original

3. Create ideas prompted by the anti-system (Nine Windows).

Let's take the example of Nine Windows for a car (short terms +/1 5 minutes) that we created previously that is shown above. Any group of Nine Windows could be used. Creating the anti-system Nine Windows for a specific problem can also prompt ideas for how to solve it.

Example

1. Create **Nine Windows** diagram.

	Past	Present	Future
Supersystem	person walking towards car, road unchanged, gas station unchanged, traffic without my car	person , road, gasoline station, traffic with my car	walking from car, road worn, gas station closed at night, traffic quiet at night
System	parked	Car	parked
Subsystem	cold engine, cold tires, cold seat, cold steering wheel	engine, tires, gas, body, seat, steering wheel	warm engine, worn tires, pollution, used gas, warm seat, moved steering wheel.

2. A second anti-system Nine Windows diagram is created in which each item in each window provides the opposite function to the original.

Anti-system Nine Windows

	Past	Present	Future
Supersystem	person walks away from car, road unchanged, gas station traffic with (holds) my car	person that doesn't control car, road that does not direct car, gas station that gathers fuel,	person walks to car, road that does not wear, gas stations open at night, traffic that produces noise at night
System	car not parked	anti-car: car that does not move passengers.	Car not parked
Subsystem	hot engine, hot tires, hot seat, hot steering wheel	engine that prevents movement, tires that don't move, gas that consumes, body that does not hold passengers, seat that does not hold passenger, steering wheel that does not direct the car	cold engine, tires that grow instead of wear, a car that cleans air rather than pollutes a steering wheel that does not direct the car.

3. Create ideas prompted by the anti-system Nine-Windows.

In this example, the anti-system prompts several radical ideas like, no travel – virtual commuting to work instead of physically being there, moving roads, tires that don't wear (floating wheels), or tires that grow by accumulating material from the road or environment, cars that do not need to be parked – someone else uses and shares it or there is mobile car parking – the car continues to move in traffic without the driver, steering wheels that turn themselves, cars that clean the atmosphere instead of pollute it, cars that pre-heat before driving them, etc.

4. The DTC Operator

The Dimension Time Cost Operator is also known as the Distance Time Cost Operator and Size Time Cost Operator.

DTC Operator is a tool for seeing the problem differently by considering it at extremes of size time and cost. What creative ideas are released by asking what would happen if the size of the system was very small, almost zero, or very large almost infinite, if the time was almost instantaneous (high speed) or very long (low speed), or the cost was almost zero or almost infinite? The DTC Operator can often release psychological inertia.

4.1 The DTC Operator Algorithm

1. Define problem: name the system or the part of system of interest.
2. Consider ideas created by DTC extremes:

Dimension:
If dimensions were extremely large, how could the problem be solved/system developed?
List ideal/solutions:

If dimensions were extremely small, (almost gone) how could the problem be solved/system developed?
List ideas/solutions:

Time:
If time were extremely long, how could the problem be solved/system developed? (E.g. days/years instead of seconds)
(Or if speed were extremely slow what would change?)
List ideas/solutions:

If time were extremely small, how could the problem be solved/system developed? (E.g. microseconds instead of seconds)
(Or if speed were extremely slow what would change?)
List ideas/solutions:

Cost: (not just in terms of dollars but cost in terms of downside, harmful effects, etc.)
If cost were extremely large, how could the problem be solved/system developed?
List ideas/solutions:

If cost were extremely small, how could the problem be solved/system developed?
List ideas/solutions

Example 1: Develop a better pen (general improvement Type 3 problem)

Dimension:
If dimensions were extremely big, how could the system be developed?

List ideas/solutions: If the pen was really long then it would be too big to carry. Can we make it flexible, wrap it around a person, and wear as a tie?

If dimensions were extremely small (almost gone) how could the system be developed?

List ideas/solutions: If the pen were really small it would become a point. A pen could be like a point on the end of a finger, perhaps ink supplied by a flexible membrane.

Time
If time were extremely long, how could the system be developed?
Or if speed were extremely slow, how could the system be developed?

List ideas/solutions: If the speed of writing were very slow – no improvement ideas created.

If time were extremely small how could the problem be solved?
Or if speed were extremely fast, how could the problem be solved?

List ideas/solutions: If the speed of writing is fast, ink would run out quickly. Develop an ink indicator, refillable pen.

Cost:
If cost were extremely large, how could the problem be solved?

List ideas/solutions: If there was no limit to cost – make a pen that writes by itself. A Dictaphone that prints.

If costs were extremely small, how could the problem be solved?

List ideas/solutions: If the cost is limited to almost zero, use free materials to write. Use the moisture in the air to mark or scrape the marks with a point. Burn the paper with the pen rather than leave ink, an "inkless pen."

Key ideas are: a pen that burns marks onto paper (heat, laser), a flexible wrap around pen, a finger pen.

Example 2: Pencil lead breaks intermittently (specific Type 2 problem)

Dimension:
If dimensions were extremely large how can the problem be solved?

List ideas/solutions: If the pencil lead were really wide then the lead would be strong and would not break? Use a very wide blunt pencil.

If dimensions were extremely small (almost gone) how could the problem be solved?

List ideas/solutions: If the pencil lead were really short then the lead would be strong and it would not break. Create a pencil with very short lead that continuously supplies lead. Propel lead from the pencil. Make the pencil recede at the same rate as the lead erodes.

Time
If time were extremely long, how can the problem be solved?
Or if speed were extremely slow how can the problem be solved?

List ideas/solutions: If time were extremely long, the lead could be protected with a cover tube for each stroke to limit stress.

If time were extremely small how can the problem be solved?
Or if speed were extremely fast, how can the problem be solved?

List ideas/solutions: If time were very short, a replacement pencil would appear as soon as breakage occurs.

Cost:
If cost were extremely large, how can the problem be solved?

List ideas/solutions: If costs were unlimited, then many sharpened pencils would be available. Use a propelling pencil (multiple sharpened points inside).

If cost were extremely small, how can the problem be solved?

List ideas/solutions: A very short pencil lead would not break, but it would not last long due to erosion.

Solutions:
A propelling pencil – with protective, retractable tubes (around the lead).

Chapter 6

Functionality, Functional Modeling and Trimming

Chapter Contents

Chapter Summary: *Functional modeling and Su-field modeling are closely related. A function statement describes **what** "thing" A does to "thing" B and a Su-field statement describes **how** "thing" A acts on "thing" B by identifying the field that operates between A and B. Function statement and Su-field are two interchangeable ways of describing an interaction.*

The quality of the interaction between A and B can be defined as useful, harmful, insufficient, excessive or absent. By combining together individual interactions (function statements) between components, functional modeling can be used to describe how a technical system or technical process operates or it can be used to model a problem situation by modeling the interactions between the components or objects involved with the problem.

Like drawing, functional modeling is a very powerful tool for releasing psychological inertia. Instead of drawing an image of what components physically look like, the image shows how components operate on each other.

Functional modeling is a powerful tool for general proactive problems (Type 3) and is useful for assessing and auditing existing technical systems and technical processes because it often reveals many problems and inefficiencies previously unrecognized. It is also a powerful tool for modeling specific problems (Type 2). By ensuring the problem interaction is included in a functional model of the technical system or technical process in which the problem resides we can identify many new potential solutions that otherwise may be overlooked.

A functional model is created graphically or in table (matrix) form. We recommend the user completes both in order to ensure all interactions are captured. Functional modeling also allows for trimming solutions where problems are solved (or a system is made more ideal) by removing one or more components and having another available component (or components) perform its useful function (or functions).

Each interaction (function statement) in a functional model can be treated like an individual Su-field. The 76 inventive standards may be used to try to find solutions to each interaction (Su-field) if they are identified as a problem or require development.

1. Functional Modeling and Su-field Modeling Introduced

There are two main ways of modeling interactions in TRIZ: Su-field modeling and functional modeling. In this chapter we will describe functional modeling and discuss Su-field modeling in detail in Chapter 8.

Functional modeling (also known as functional analysis) originated from Value Engineering. It was first developed by Larry Miles in the 1950's. It is a data gathering tool that captures information about the functional relationships between components in a technical system or process. It is recommended for use with a Type 2 specific problem and for a Type 3 general inventive problem. But because it is also useful as a data gathering tool it may be used for all four problem types. It is useful for gathering information for failure prevention (even though it is not a failure prevention tool) and for gathering operational information for a technical system or technical process with unknown root cause.

Before we discuss functional modeling it is useful to understand the differences between functional modeling and Su-field modeling because the two are closely related. Both functional and Su-field models describe interactions. A function statement is a single interaction. A functional model can be many interactions and include more than two components. A Su-field model describes a single interaction between two objects, similar to a function statement. A basic interaction between two objects can be stated in terms of Subject – Action – Object. The subject performs an action on an object where the action is a verb. This is also the common structure of sentences in language, Subject – Verb – Object or "A" acts on "B", where "A" is the subject and "B" is the Object.

For example:

Horse – pulls – cart
Sun – heats – air
Cup – holds – coffee
Gravity – moves – water

When the subject performs an action on the object, one or more parameters of the object is maintained or changed by the action. In the above examples, the position of the cart changes, the temperature of the air changes, the position of the coffee is maintained, the location of the water changes.

This basic interaction is represented as a subject and object that are connected by an arrow; the direction of the connecting arrow indicates the direction of the action.

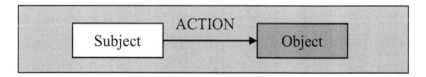

Or simply "A" acts on "B."

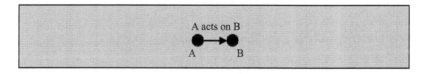

We define the *quality* of an interaction by its effectiveness. We use the terms useful, harmful, insufficient, excessive, or absent to describe the quality of an action.

USEFUL: the action is desired and not considered to be a problem. The action is effective.
HARMFUL: the action is undesired. It does "something" negative.
INSUFFICIENT: the action is useful but the action is less than is needed to deliver the action to the desired level.
EXCESSIVE: the action is useful but is greater than is needed.
ABSENT: there is no interaction between the subject and object

The convention commonly used is to illustrate these actions is denoted by the different types of connecting arrows shown below. "Absent" has no connector.

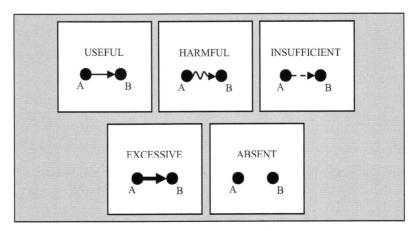

Note that an absent action may also be drawn as an action with an "X" through it. It has the same meaning as no connector, i.e. "A" does not act upon "B." This X notation is often used to emphasize that there is a desired action by virtue of being absent. For example, blunt pencil does not pierce paper may be a desired action, (this is a null positive action, we will discuss these later in Chapter 9, Contradictions and ARIZ tools).

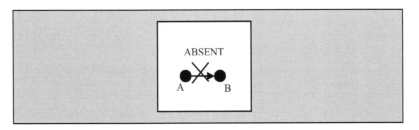

Useful, harmful, insufficient, excessive or absent are a qualitative assessment of the interaction, they represent an opinion. For example, let's look at the action of inserting a nail into a wall, depending on circumstance (the subject, the object, the force) the quality of the action can be assessed in any of the five ways.

If we use a standard claw-hammer to insert a nail then the interaction "hammer – moves – nail" is defined as useful. If we use a folded newspaper to try to drive a nail into the wall then the nail barely moves. The function, "paper – moves –nail" would be described as insufficient.   If you use a 10lb sledgehammer that shoots the nail far into the wall then the interaction would be excessive. If by accident you hit your thumb with the hammer then we create the statement "hammer – moves – thumb" which we can denote as harmful. If you don't hit the nail at all, the interaction is absent.

Although both function statement and Su-field models describe the basic interaction of subject-action-object, or A acts on B, functional modeling describes **what** interaction takes place. Su-field modeling describes **how** the interaction takes place.

Function Statement: A – function – B. A acts on B via a function (Fn).

- A and B can be any "thing", any physical object, substance or a field.

Su-field Statement: S2 – field (F) – S1. S2 acts on S1 via a field (F)

- A and B (or S2 and S1) can be any physical object or substance, but NOT a field

Function statements and Su-fields are two ways of representing the same subject – action – object (SAO) interaction.

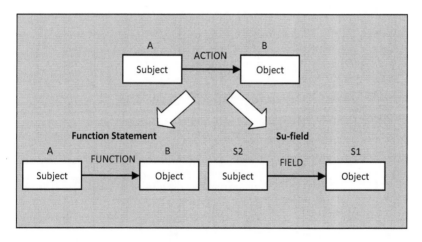

Function statements and Su-field interactions are drawn differently (in fact there are a few conventions for drawing a Su-field which we will discuss in Chapter 8 on Su-field modeling - we will use the triangle barbell notation).

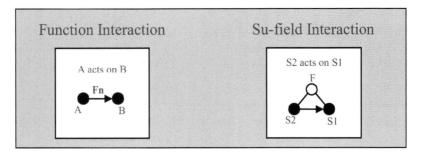

Note that in Su-field modeling the convention is to use S2 and S1 rather than A and B or subject and object. S2 and S1 are abbreviations for substance 2 and substance 1. The term substance is not a good translation from Russian since substance in English implies a "material" like iron, wood, plastic, etc. S2 and S1 can be any physical object (person, chair, sand etc) but not a "field." A better translation is "thing."

A function statement tells you what A does to B. The function like an action is a verb. For example: drill removes wood.

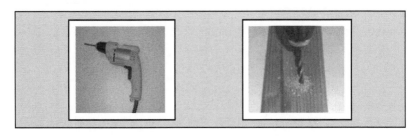

A Su-field statement tells you how A acts on B. Only the field is described. In this example: drill operates on the wood via a mechanical field.

Both function statement and Su-field statements could be combined to say: A performs function "Fn" on B by using field F. For example the drill removes wood by using a mechanical field.

horse – pulls – cart
sun – heats – air
cup – holds – coffee
gravity – moves – water

An **action** is not always the same as a function. An action can be made up of several basic functions. In the examples above, horse pulls cart, sun heats air etc, the action is the function, but in the statement "man cooks egg" the action word cooks is made up of several functions. Man moves spatula, spatula moves egg, flame heats pan, pan heats oil, oil heats egg. Cooks is not an acceptable function description. When we discuss functional modeling below

98

we will discuss how to create the correct functional language that defines basic single functions that make up actions.

Above we stated that for Su-field modeling A and B can be any "thing" but not a "field." Let's explain this further. A "thing" (A or B) is allowed to be a field in functional modeling and not a field in Su-field modeling. This is because in Su-field modeling the interaction requires showing two physical entities interacting via a field.

The four statements (horse pulls cart, etc.) were function statements. Let's see what they look like expressed as Su-fields and combined function-Su-field statements to illustrate the difference between Su-fields and function statements.

Function Statement	Su-field	Combined
A Fn B	A Field B	A Fn B via Field F
horse pulls cart	Horse acts on cart via a mechanical field	Horse pulls cart via a mechanical field
sun heats air	Sun acts on air via a thermal field	Sun heats air via a thermal field
cup holds coffee	Cup acts on coffee via a mechanical field	Cup holds coffee via a mechanical field
gravity moves water	Earth acts on water via a gravitational field	Earth moves water via a gravitational field

Note that in the last example, gravity is a field. Therefore we can't use it as the subject for a Su-field statement. We must find the physical entity that generates the field. In this case the planet Earth and the water are attracted via a gravitational field.

A function statement can be converted to a Su-field statement and vice versa. Given a function statement, we can create a Su-field, allowing access for how to solve a Su-field problem (using the 76 inventive standards). A Su-field can be converted to a function statement by identifying the function that acts between objects, allowing access to methods of solving function statements (using the scientific effects database see Chapter 7).

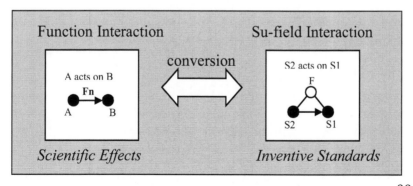

By combining multiple function statements, functional modeling can be used to model entire technical systems or technical processes. Systems and processes can be broken down into their individual interactions and process steps then qualitatively assessed for effectiveness. Functional modeling releases psychological inertia in a similar way to drawing a system or process. We can see all the parts and how they interact between each other and with the outside world. This helps us to see the problem situation differently.

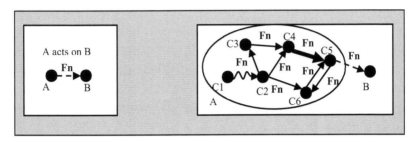

A functional model of a technical system is made up of one or more function statements and each one can be converted to a Su-field; each Su-field is evaluated for the quality of its interaction (useful, harmful, insufficient, excessive, absent) and thereby we define the interaction as a problem. A harmful interaction should be eliminated, insufficient strengthened, excessive weakened and absent made present if needed. Each interaction problem can be solved individually. As a Su-field statement we can use the 76 inventive standards which is a list of generic solutions for eliminating a harmful interaction, strengthening a weak interaction, etc. To address an individual function statement we can search the scientific effect database for alternative ways to deliver the function.

Building *accurate* function statements is necessary for effective functional modeling. Using simple language releases psychological inertia. This means we use accurate simple language to describe functions called functional language.

2. Functional Language

Functional language is the use of simple words that accurately describe the direct physical action that A is performing on B to change or maintain one or more parameters of B in a function statement. We recommend choosing the minimum number of words (preferably one) to describe the function.

High level complex or technical language is a carrier of psychological inertia. By using simple language to describe the function performed by a system or component we release psychological inertia – it becomes easier to think of alternative ways of providing that function. The simpler language prompts us to consider alternative scientific effects as innovative solutions or to find different ways of performing the function using available resources which can lead to innovative design changes and solutions.

2.1 Function Statement Examples

In the picture below, what is the function statement (Subject-Function-Object) that describes the action between the shelf and books?

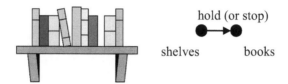

Function statement: shelf holds (or stops) books.
The function hold (or stops) is applied by the shelves to the books. The parameter of the book that is changed or maintained by the shelves is position, even though the books do not move. If the shelves were not present the books would fall. The parameter position is maintained by the shelves. Non-functional language, language that does not accurately describe the function would be shelves-contain-books.

What is the function statement between knife and cheese?

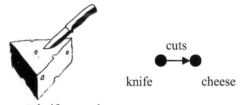

Function statement: knife cuts cheese.
The knife is the subject. It performs the action "cuts" on the object, cheese. The parameter of cheese that changed is shape or level of segmentation.

What is the function statement between the person and the computer?

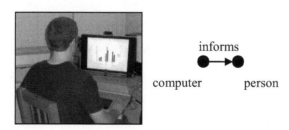

Function statement: computer informs person."Informs" is the function the computer delivers to the person. The level of information is the parameter of

101

the person that is changed. Person views computer would not be a correct function statement.

In a function statement, the action must change a parameter of the recipient (subject acts on object). The function description must be simple and physically immediately direct.

What is the function statement between the umbrella and rain?

Function statement: umbrella stops rain.
We use umbrella stops rain not umbrella protects person. "Protects" is a general description of the function, there is no parameter of rain that is changed by "protects." The parameters of rain (object) that are changed by stops are speed, shape, acceleration, direction, etc.

What is the function statement between the cup and coffee?

Function statement: cup holds (or stops) coffee.
Although cup–contains–coffee is correct in everyday language, the word "contains" is descriptive, not simple and direct, it is therefore not useful for the simple functional language we use for functional modeling.

A sack of potatoes can contain air, but it does not stop it. The air flows through the sack cloth. Sack stops or holds potatoes would be correct using functional language. We need air flow to preserve the potatoes freshness for eating. So we could add a second function, sack passes air.

What is the function statement between the broom and dirt?

moves

broom   dirt

Function statement: broom moves (or removes) dirt.
Not broom – cleans - floor ("cleans" is descriptive but not simple and direct) or broom – brushes - floor which is also unclear functional language. "Moves" is the correct functional language, the parameter, location of the dirt is changed by the broom moving. There is no simple parameter of the dirt changed by "cleans" or "brushes."

Person – fries – food is not simple direct functional language. To "fry" contains many functions. It requires utensils to be moved, heat to be adjusted, food to be manipulated. It is not a single direct function but a collection of functions.

It is important when building function statements that the subject and object are not measurements of parameters. For example, ice – cools – temperature. Pressure – moves – gage. Weight – moves – plank. Temperature, pressure and weight are parameters not physical objects, fields or forces and therefore they should not be used for function statements.

What is the function statement between the bottle and ice?

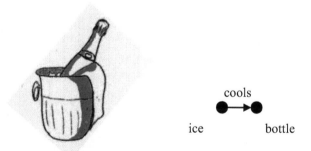

cools

ice   bottle

Function statement: ice cools bottle.
We can treat fields and forces (gravity, magnetic field, etc.) as objects when we construct functional statements (as we discussed earlier, this is not the case for Su-field modeling). For example, gravity – holds – chair.

Vacuum is a special case that can be treated like an object, even though it is not an object but a parameter (low pressure, absence of gas). For example, vacuum cups are used to lift sheets of glass.

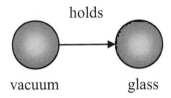

Function statement: vacuum holds glass

For a special type of subject that generates a field, the function "emits" can be used. For example, magnet – emits – magnetic field. Laser – emits – light.

What is the function statement between the magnet and nails?

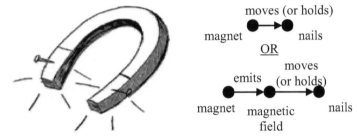

"Magnet - moves - nails" or "magnet emits magnetic field" combined with "magnetic field moves nails" is acceptable in functional modeling. However, usually we use the most basic, which in this case is "magnet moves nails" or "magnet – holds - nails." When we learn Su-field modeling, fields are not considered as objects, we would say magnet holds nails by means of a magnetic field.

Important rules to remember when building function statements:

- A function statement defines the interaction that occurs between two objects in the form subject-action-object.
- The subject must be the provider of the action, the object the receiver and a parameter of the receiver must be changed or maintained by the function.
- Do not use a parameter as a subject or object. For example, do not use temperature - heats - water.
- Use simple basic non-technical language, simple, not descriptive verbs to describe the function that directly links one component to another.
- Fields can be treated as subjects or objects in function statements.
- We can treat a vacuum as a subject or an object in functional modeling.

Exercise:

Some examples (examples with strikethrough are incorrect), explain why.

Subject	Function	Object
Bottle	~~contains~~	Water
Bottle	holds	Water
Bottle Cap	~~seals~~	Bottle
Bottle cap	stops	Liquid
Device	~~measures~~	Signal
Signal	informs	Device
Lightning rod	~~protects~~	Building
Lightning rod	conducts	Current
Brush	~~cleans~~	Floor
Brush	removes	Dirt
~~Temperature~~	heats	Water
Lamp	heats	Water
Mirror	informs	Person
Light	~~improves~~	~~Vision~~
Light	informs	Person
Lens	~~Magnifies~~	Object
Lens	bends or focuses	Light
CD Player	~~Plays~~	~~Music~~
CD Player	Informs	Person
Person	~~drives~~	Car
Car	transports	Driver
Pipe	~~moves~~	Water
Pipe	directs	Water

Before we discuss how to build a functional model of a technical system, we need to define the main useful function (MUF) of a technical system and the product of a technical system.

2.2 Main Useful Function (MUF)

A technical system is a system that is made up of one or more element that operates to perform or provide one or more functions. Examples of technical systems are a pen, ruler, train, computer, microscope, airplane etc. Non-"manufactured" systems can be modeled as a technical system, for example, a cow can be modeled as a technical system even though it wasn't "manufactured."

The main useful function of a technical system is the main function or functions that the technical system performs. Think of it as the purpose or purposes for which the technical system was made.

The main useful function (MUF) of a car is to move people (or a person). Car – moves - person. The main useful function of a lawnmower is to cut grass.

Lawnmower – cuts - grass. The main useful function of a book is to inform people (or person). Book – informs - person.

What is the main useful function of an electric kettle?

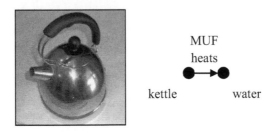

It is to heat water.

Function Statement: kettle heats water.

What is the main useful function of a window? Windows are made to allow light in (to direct or pass light), and to stop the environment from outside entering the room. It has two main useful functions.

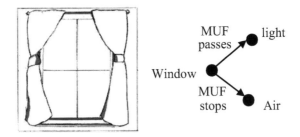

Function statements: window passes (directs and transmits are acceptable alternatives) light, and window stops air.

In the picture below, what is the main useful function of a mirror? Is it to reflect light, or to inform person? Although the mirror performs the function "to reflect light", the purpose of a mirror is to inform a person.

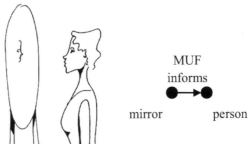

If the mirror were used inside a laser cutting machine that is used to cut objects using laser light, the main useful function of the mirror inside the machine would be to reflect light or direct light. The main useful function is dependent upon the purpose(s) for which the technical system or component was made.

2.3 Product Definition

Technical systems are made up of one or more components designed to perform a specific function or functions. The main useful function of a technical system is performed on an object or objects. The general convention is to call the object (or objects) of the main useful function statement the **product** (or products) of the system.

In functional modeling the product of a car is a person. The main useful function statement is: car – moves - person. Person is the object of the main useful function and therefore we call the person the product of the technical system. We also know that a car produces pollution and pollution is a product of the car. However, producing pollution is not the main useful function of a car and in functional modeling the person is the product of a car.

If the technical system is a lawnmower, the product is grass. Grass is the object in the main useful function statement, lawnmower - cuts - grass.

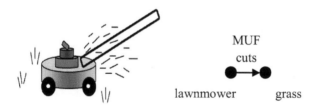

A system can have more than one product. In the example of a window above, the products are air and light. A product is defined as the object (or objects) of the main useful function (or functions) of a system.

3. Building a Functional Model of a Technical System

Above we learned how to build accurate function statements. Technical systems are usually made up of many function statements combining many components and interactions. The diagram below shows a situation where, technical system "A" has one component, two components, and multiple components.

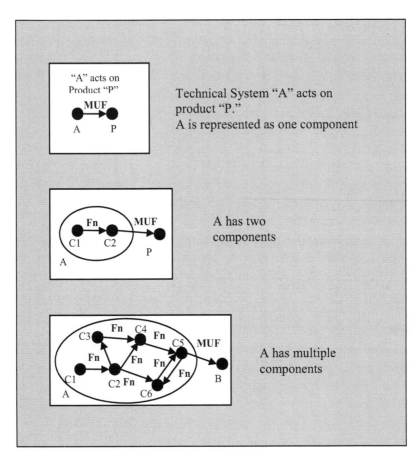

"A" acts on Product "P"

MUF

A P

Technical System "A" acts on product "P."
A is represented as one component

Fn **MUF**

C1 C2

P

A

A has two components

Fn C4 **Fn** C5 **MUF**

C3

Fn **Fn** **Fn**

C1 C2 **Fn** **Fn**

C6

A

B

A has multiple components

A bat has a handle and "head." A bottle has a body, neck, cork, seal, etc. We call these elements that make up a technical system the *components*. They are individual parts of a technical system; they do not necessarily have to be physically separated. A comb has a handle and teeth. They are joined together but are separate components in a functional model.

System components (C1, C2, etc.) also interact with external components, the outside world, the air, gravity, nearby objects and so on. For example, gravity moves arm; air cools pipe.

External components are known as the supersystem (SS1, SS2, etc.). When building a functional model of a technical system, we add supersystem components that our system interacts with.

By adding actions and the description (useful, harmful, insufficient, excessive and absent) of the quality of actions between components of the system, supersystem and product we can build a functional model the system.

3.1 Graphical Format

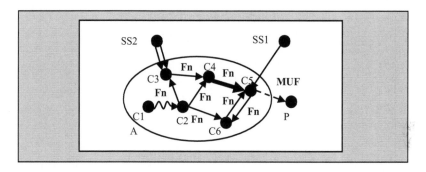

This is an example of a functional model of a technical system in graphical format. It is a powerful tool for releasing psychological inertia. It is useful for identifying problems with a technical system that may otherwise have been overlooked, interactions that may not have been considered to be a problem are often revealed.

Below we will discuss how to "trim" a functional model, trimming can be used to try to improve a technical system or process by making it more efficient and to identify potential solutions by removing rather than adding components. Functional modeling is also a useful data gathering tool; it helps describe how a system functions.

To be manageable it is important to limit the number of components that are included in the model. The general recommendation is to try to keep the number of components at or below ten, including supersystem and product components.

Try to include only the main components that perform the key functions which describe the main useful function of the system. Do not include every part, screw, cooling line, etc. Combine parts into single subsystem components, if necessary and build separate models for each subsystem.

There is no single "correct" functional model. There is an element of interpretation required by the user. A functional model of the same system created separately by two people will normally produce two different models. The components chosen will often be different and quality of interactions may not match. Differences are to be expected, the functional model is a tool for thinking not an engineering blueprint.

The functional model is an analytical tool to help the user think by laying out the components and functions in a way that is easily analyzed rather than trying to think only in their head. Like a map or drawing it enables the user to see the system in a way that is easier to follow and manipulate.

3.2 Matrix Format

Often users like to build a functional model of a technical system graphically as in the figure above. But many prefer simply building a matrix. The matrix table for the graphical model above is given below. The information contained in both formats is the same. Both can be used in parallel. They complement each other. Often if only the graphical model is created, some interactions between components are missed by the user. The matrix table helps identify interactions and ensure none were missed. The graphical representation helps visualize trimming solutions more easily than the matrix format.

In the matrix format of the functional model, the component in the first column "acts on" the components in the subsequent columns to the right. The action describes the function (using simple language) and quality of the interaction. For example C1 heats (harmfully) C2, which matches the graphical representation shown above.

Complete Functional Model of a Technical System (matrix format)

	C1	C2	C3	C4	C5	C6	SS1	SS2	Product
C1		Heats (H)							
C2			Holds (U)	Moves (U)		Removes (U)			
C3				Rotates (U)					
C4					Cools (E)				
C5						Stops (U)			Informs (I)
C6					Moves (U)				
SS1					Holds (U)				
SS2			Holds (U) Moves (U)						
Product									

When the functional model has been created and the problem interactions identified, interactions are selected for improvement using solutions tools.

3.3 Steps for building a Functional Model of a Technical System

1. Form the main useful function statement

This will identify the technical system (TS), main useful function (MUF) and product(s) (P).

Examples are drill (TS) removes (MUF) wood (P), kettle (TS) heats (MUF) water (P), bulb (TS) illuminates (MUF) book (P).

Main useful function

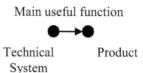

Technical　　　　Product
System

2. List the main components of the technical system.
3. List the main components of the supersystem.
4. In parallel
       a. Build a graphical model including interactions and their quality.
       b. Build a matrix model.
5. Identify the interactions that should be improved.

3.4 Examples of Technical System Functional Models

When we build a functional model we describe the functions of each component and the component(s) it affects. We do not consider time. We will consider time when we discuss functional modeling of a technical process below. As a simple example of how to build a functional model, let's make a functional model of a car.

Functional Model of a Car

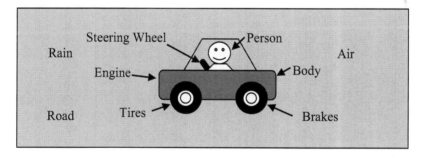

1. Main useful function statement: Car – Moves - Person

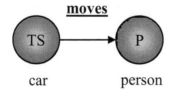

car – moves - person

    Technical system: car
    Main useful function: moves (or transports)
    Product: person

111

2. List components of the system.
   Make an Inventory of all the components of the system.
   Main components are: C1 tires, C2 body, C3 engine, C4 steering wheel and C5 brakes.

3. List components of the supersystem
   Make an inventory of the external elements that are not designed as part of the system but can interact with it.
   In this case we choose: SS1 road, SS2 rain, SS3 air. We could choose to add pollution, traffic, etc. but won't in this case.

4. In parallel, build a graphical and matrix functional model

   a) Build a functional model of the car (graphical format).

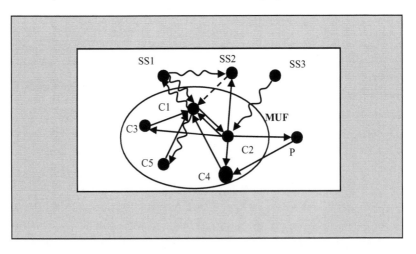

   b) Build a functional model of a car (matrix format)

	C1 Tires	C2 Body	C3 Engine	C4 Steering Wheel	C5 Brakes	SS1 Road	SS2 Rain	SS3Air	P Person
C1 Tires		Move (U)			Erode(H)	Erode(H)			
C2 Body	Holds(U)		Holds(U)	Holds(U)			Stops(U)		Moves(U)
C3 Engine	Rotates(U)								
C4 Steering Wheel	Directs(U)								
C5 Brakes	Stop(U)								
SS1 Road	Erodes(H)						Holds(H)		
SS2 Rain	Holds(I)								
SS3 Air		Stops (H)							
Product Person				Directs(U)					

Tires move body, erode brakes erode road (remove is simpler functional language than erode).

Body holds tires, holds engine, holds steering wheel, stops rain, and moves person (main useful function).
Engine rotates tires.
Steering wheel directs tires.
Brakes stops tires.
Road erodes (or removes) tires (removes rubber from the tires to wear them away). Road holds rain.
Rain holds tires (when the road is wet holding is insufficient, tires often skid).
Air stops body (retards the progress of the car).
Person directs steering wheel.

We have defined a number of problems with the car. We can convert these problems into function or Su-field problem statements (in the form of how to: improve or perform an interaction).

5. Identify the interactions that should be improved.

Function Problem Statement	Problem
Road erodes tires (harmful)	How to: remove the function erodes or removes?
Tires erode road (harmful)	How to: remove the function erodes or removes?
Road holds rain (harmful)	How to: remove rain that is held? How to remove rain?
Rain holds tires (insufficient)	How to: improve the function holds?
Air stops body (harmful)	How to eliminate the function stops?
Tires erode brakes (harmful)	How to eliminate the function erodes?

Note: we could try to turn these into any of the classical TRIZ problems. For example, the first example can be converted to the contradiction: if the tires are hard then they do not erode but they do not grip the road (technical contradiction). Or the tires must be hard to not erode and soft to grip the road, the tires must be hard and soft (physical contradiction). We discuss the reformulation of specific problems into contradictions, Su-fields and function statements and how to solve them in subsequent chapters.

The above was an example of how to use functional modeling for a general non-specific (Type 3) inventive goal. We can also use functional modeling for a specific goal (Type 2) problem.

3.5 Using Functional Modeling for Specific Problems

When the root cause of a problem is known it is important to try to include it as a function interaction in the functional model. For example, if there is a leaking pipe, include pipe - holds - water insufficiently in the model. If parts are being broken in a technical system, include the function that causes it in the model; arm - transports - part harmfully.

If the cause of a specific problem is unknown, for example you do not know where unwanted breakages are coming from, then you cannot include root cause in the model. Functional modeling does not identify root cause. Building a functional model can help collect information about the system in a way that often releases psychological inertia. We see the system differently and this often prompts new ideas of potential causes (see Chapter 11). If it is a complex system with many components, try to focus on the area where the problem is believed to be occurring.

4. Trimming

In addition to evaluating a system to identify problem interactions, a completed functional model of a technical system provides a trimming opportunity. Trimming is a method of simplifying a system by removing components or parts of components in a system but maintaining functionality. It can therefore be used for improving the ideality of a technical system (Type 3 general inventive goal).

Trimming may also solve specific (Type 2) problems. We have a tendency to add components to solve problems. Rather than adding components to solve a specific problem, trimming can solve problems by removing components which is counterintuitive.

4.1 Examples of Trimming

Example 1: Freezer

A freezer contains food that you want to keep frozen. Occasionally the door is left open causing the food to defrost. The problem is to stop unwanted defrosting of the food. List your own ideas before we apply functional modeling.

Typical initial brainstormed ideas are:

- Attach a proximity alarm to the door. It will sound when the door is open and raise the alarm to close the door.
- Place a mechanism inside the freezer that detects light and alarms when the door is open.
- Install a mechanical door closing device.
- Put a large notice on the door to remind people to close the door.
- Place a bungee cord around the door and freezer.

First build a functional model of the system. When building a functional model of a system to address a specific problem with known root cause, it is important to define the system as it normally functions but also to include the specific components and interactions that describe the problem. In this case a

114

person only sometimes closes the door, we will name this function as "person insufficiently closes door."

1. Define the main useful function statement
Freezer – cools - food

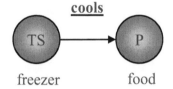

freezer   food

2. Make an inventory of all the components of the system.
Main components are: body, motor, door, hinges.

3. Make an inventory of the external elements (supersystem) that are not designed as part of the system but can interact with it.
Supersystem components are: SS1 air, SS2 person, SS3 gravity.

4. Build a graphical and matrix functional model

Functional Model of Freezer (graphical format)

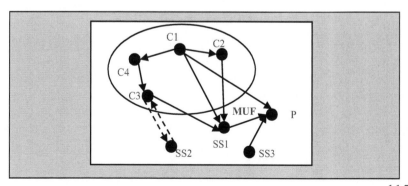

Functional Model of Freezer (matrix format)

	C1 Body	C2 Coils	C3 Door	C4 Hinges	SS1 Air	SS2 Person	SS3 Gravity	Product Food
C1 Body		Holds(U)		Holds(U)	Stops(U)			Holds(U)
C2 Coils					Cool (U)			
C3 Door					Stops(U)	Informs(I)		
C4 Hinges			Hold(U)					
SS1 Air								Cools (MUF)
SS2 Person			Moves(I)					
SS3 Gravity								Holds(U)
Product Food								

Body holds coils, holds hinges, stops air and holds food.
Coils cool air.
Door stops air and informs person insufficiently that the door is not closed.
Hinges hold (direct) door.
Air cools food (main useful function).
Person moves door insufficiently, not enough to close the door repeatedly.
Gravity holds food (gravity holds the food on the shelves).

5. List the interactions for improvement (the problems).

Function Problem Statement	Problem
Door informs person (insufficient)	How to improve door informs person?
Person moves door (insufficient)	How to ensure the person always closes the door?

In subsequent chapters we will learn how to find a solution to improve these Su-field interactions/ function statements using inventive standards or find an alternative way of performing them (research scientific effects for how others have solved the problem in the past). Or we can consider trimming the problem function(s).

Let's try trimming the insufficient function person – moves - door. If we remove the component person then we need to ensure we still provide the function to move door. The simplest component to transfer the function to is gravity (SS3) in the supersystem. If we transfer the function "moves door" to gravity we no longer need to inform the person, so we can delete the function door - informs - person.

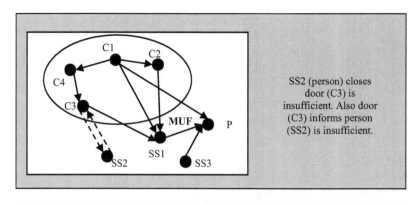

SS2 (person) closes door (C3) is insufficient. Also door (C3) informs person (SS2) is insufficient.

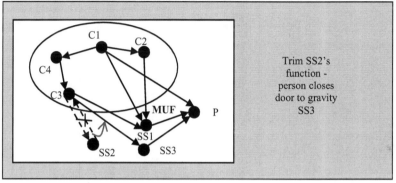

Trim SS2's function - person closes door to gravity SS3

How to provide the Function: Gravity - moves (closes) – door?

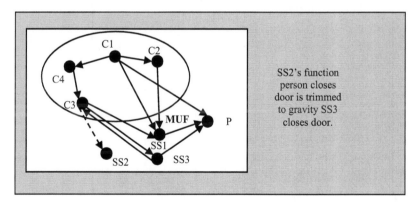

SS2's function person closes door is trimmed to gravity SS3 closes door.

Now we have created the trimming problem we can think on ways to solve it. How can gravity close the door?

If we tilt the freezer back by lengthening the front legs slightly, then gravity will automatically close the door!

117

If the door closes by itself, we can also remove the function door insufficiently informs person.

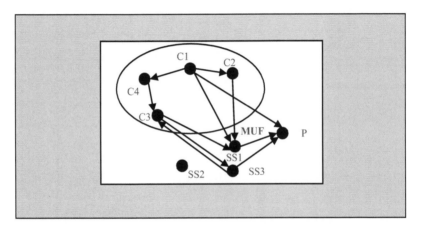

Often it is simpler to solve a trimming problem than the initial problem we define. Trimming can be used to create many innovative ideas even for very complex engineering problems. Trimming is an extremely useful tool for removal of psychological inertia. Functional modeling is a useful tool for helping us see trimming solutions. We can find many simple highly inventive solutions by removing rather than adding, the initial typical brainstormed ideas shown above illustrates our tendency to find solutions by adding to the system rather than seeking simpler more ideal trimming solutions.

Example 2: Plasma Etch

High energy ions are accelerated towards a silicon wafer as part of the manufacturing process for semiconductor devices. An unwanted deposit of material builds up on the edge of the electrode that eventually impacts performance. How do we remove the build-up? Add a protective shield, add

118

special cleaning steps? The simple trimming solution is to remove the deposition location without impacting performance.

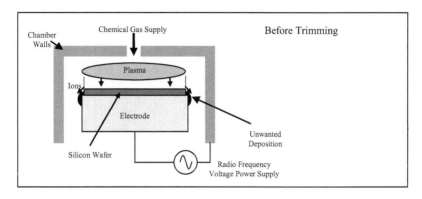

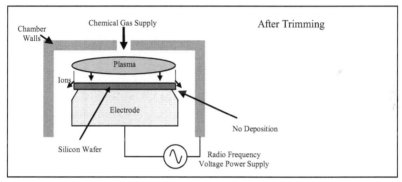

Of course when you trim, you can't just delete objects or move functions. You may have to replace useful functions that are deleted when a component is removed. In this case, the edge of the electrode performs no useful function.

By creating a functional model and experimenting with the removal of components (or parts of them) and considering how the system would be re-designed is a powerful way of prompting innovative ideas for system improvement.

5. Algorithm for Functional Modeling and Trimming of a Technical System

Step 1 Main useful function statement
Form the main useful function statement.
For a specific problem, create a function statement that describes the problem, and include the interaction in the functional model of the system.
Step 2 Make an inventory of all the components of the system.
Main components are: C1, C2, C3, etc.

119

Step 3 Make an inventory of the supersystem elements

Make an inventory of external elements that are not designed as part of the system but can interact with it. Supersystem components are labeled as: SS1, SS2, SS3 etc.

Step 4 Build a graphical and matrix functional model.

Step 5 List the interactions for improvement.

Solve the problems using personal knowledge. Solve the function statement using scientific effects. Convert the function statement to a Su-field and apply inventive standards. We will discuss the scientific effects and inventive standard solutions tools in later chapters.

Step 6 Create a trimming problem.

Propose removal of one or more functions or components and remove the negative function and transfer to useful functions to other components. Solve the trimming problem. Create ideas for how to make the make a trimmed system work.

6. Functional Modeling of a Technical Process

The functional models of a technical system described above merely show what interactions take place in a technical system. There is no indication of the sequence functions are performed. The intention is simply to reveal the operation quality and provide a trimming opportunity for the interactions. In process modeling we model the actions that take place in time. The sequence of how steps are performed can also yield innovative solutions.

6.1 Building a Functional Model of a Technical Process

The same "subject – function – object" methodology used to build a functional model of a technical system is used to describe each step in a technical process. In functional modeling of a technical process, a step is the same as a subject – function – object interaction. The technical process is broken down into a series of subject-function-object interactions for a technical system but this time they are listed in sequence. If needed, the start and end time of each step can be identified. Steps performed in parallel are indicated by the start and end time.

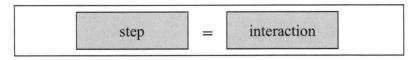

When building a functional model of a technical process, it can be useful to add cost information for each if the objective is to move towards a more ideal system. Adding a cost column can help the user target the improvement or elimination of highest cost steps. Note that, if the number of steps becomes very large, it is a good idea to break the process into phases where each phase contains multiple steps with a similar objective. A table is used to list the steps

Step	Start/End Time	Subject	Action	Object	Type	Function Description
1						
2						
Etc.						

This is a very powerful tool for *process auditing* and releasing psychological inertia. It allows us to consider the quality of each step (useful, harmful, insufficient, excessive, absent) and apply the tools for improving or changing the interaction.

Often maintenance procedures that have been performed for many years in industry are significantly improved by the release of psychological inertia that occurs simply by stating each step as an interaction. Steps previously not even considered to be a problem are revealed as an opportunity for improvement.

For example, during a preventative maintenance procedure, metal parts are left to cool for several hours in order to be handled at the next step. This was not considered to be a problem until the procedure was audited using a process functional model and the step was identified as insufficient. Instead of waiting several hours for these parts to cool, they were forced to cool quickly via a water cooling loop significantly increasing production time of the technical system.

Technicians can spend years performing the same procedural steps without questioning or realizing there are simple and highly effective improvements possible. The potential for improvement is only realized when the problem has been identified using a process functional model.

7. Algorithm of Functional Modeling and Trimming of a Technical Process
1. Define the purpose of the process.
2. Define the main phases (some processes can be broken down into stages in work type, objective, blocks of time, etc.).
3. Define the individual steps and their quality.
4. Make a list of the problem steps to improve.
5. Define an improvement plan including trimming ideas.
   a. Are there steps that are not needed?
   What steps can we do in parallel?
   b. What steps can be performed by another step?
   c. Can the sequence be improved?
   d. How to improve problem steps (use personal knowledge, scientific effects, inventive standards, the TRIZ solutions tools).

121

7.1 Example: Boiling an Egg

1. The purpose: to boil an egg.
2. Main Phases 1.Preparation 2.Boil the egg 3.Remove and plate the egg.
3. Define the individual steps and their quality.

Phase 1: Preparation
Description:
Gather the materials: get an egg from the refrigerator, get a pot from the cupboard, get a timer, get the egg-cup and tongs. Prepare the timer, place the egg in the pot, fill the pot with water, and place the pot on the stove.

Put each step in the format subject-function-object. Use simple functional language. Avoid unnecessary detail, person moves fingers, person lifts pot, person moves arm, for example is too much detail. Even though fingers are used, person moves pot is sufficient. The level of detail (number of steps to include) is subjective; too much detail will be cumbersome.

Step	Time	Subject	Function	Object	Type	Description/comment
1.1		Person	Moves	Door	Useful	Open refrigerator
1.2		Person	Moves	Egg	Useful	Take out egg
1.3		Person	Moves	Door	Useful	Close refrigerator
1.4		Person	Moves	Pot	Useful	Get pot from cupboard
1.5		Person	Moves	Egg	Useful	Place egg in pot
1.6		Person	Directs	Egg	Useful	Place egg in pot
1.7		Pot	Holds	Egg	Excessive	Egg may break when placed in pot.
1.8		Person	Directs	Pot	Useful	Place pot under faucet
1.9		Person	Rotates	Faucet	Useful	Turn on water
1.10		Faucet	Directs	Water	Useful	Water fills pot
1.11		Pot	Holds	Water	Useful	Pot is filled
1.12		Light	informs	Person	Useful	Fill to top of egg
1.13		Person	Rotates	Faucet	Useful	Close off water
1.14		Person	Directs	Pot	Insufficient	Move pot to stove
1.15		Person	Transports	Pot	Insufficient	Water may spill
1.16		Stove	Holds	pot	Useful	Pot is on stove

Phase 2: Boil the Egg
Description:
The gas (or electric power is turned on), the water is heated until boiling and boiled for three minutes.

Step	Time	Subject	Action	Object	Type	Description/comment
2.1		Person	Rotates	Knob	Useful	Turn on gas
2.2		Person	Ignites	Flame	Useful	Flame ignites
2.3		Flame	Heats	Pot	Insufficient	Transfer of heat depends on the contact of the flame with the pot, if pot is small; heat is lost around the sides. If too big, excess water is heated.
2.4		Pot	Heats	Water	Excessive	Excess water is heated. We only need enough to surround the egg.
2.5		Water	Heats	Egg	Useful	Water heats the egg
2.6		Water	Informs	Person	Useful	Person observes boiling
2.7		Person	Moves	Timer	Useful	Person starts timer
2.8		Timer	Informs	Person	Useful	Timer alarms after 3 minutes

Phase 3: Remove and Plate the Egg
Description:
Person removes the pot from the stove, removes the egg from the water with tongs, places the egg in an egg-cup; empties hot water.

Step	Time	Subject	Action	Object	Type	Description/comment
3.1		Person	Rotates	knob	Useful	Turn off gas
3.2		Person	Moves	Pot	Useful	Move pot from stove
3.3		Person	Directs	Pot	Insufficient	Water easily spilled
3.4		Person	Moves	Tongs	Useful	Direct pot to table
3.5		Tongs	Hold	Egg	Useful	Pick up egg
3.5		Person	Directs	Tongs	Useful	Move Egg with tongs
3.6		Cup	Holds	Egg	Useful	Move egg to egg-cup
3.7		Person	Moves	Pot	Useful	Empty the pot

4. Problem Steps to Improve

Step	Time	Subject	Action	Object	Type	
1.7		Pot	Holds	Egg	Excessive	Egg may break when placed in pot.
1.14		Person	Directs	Pot	Insufficient	Move pot to stove
1.15		Person	Transports	Pot	Insufficient	Water may spill
2.3		Flame	Heats	Pot	Insufficient	Transfer of heat depends on the contact of the flame with the pot, if pot is small; heat is lost around the sides. If too big, excess water is heated.
2.4		Pot	Heats	Water	Excessive	Excess water is heated. We only need enough to surround the egg.
3.3		Person	Directs	Pot	Insufficient	Water easily spilled

5. Define Improvement Plan and Trimming Ideas

   a. Steps that are not needed:
       • Steps 2.7 and 2.8 turning timer off and on. We could use a clock.
       • Step 3.7 person - moves – pot. Pot does not need to be emptied. Another egg could be cooked more quickly using the hot water in the pot.
       What steps can we do in parallel?
       • We could get the tongs while the water is boiling.
   b. What steps can be performed by another step?
       • Trim 2.7. Instead of person starts timer; have the boiling water start the timer.
   c. Can the sequence be improved?
       • We can fill the pot before placing the egg in it. Eliminating the excessive action in 1.7 that can break the egg.
       • We could fill the pot while it's on the stove using a jug to avoid spillage eliminating the insufficient action in Step 1.15.
   d. How to improve problem steps (use personal knowledge, scientific effects, inventive standards, the TRIZ solutions tools).

- 2.3 flame – heats - pot is insufficient and 2.4 water – heats - egg is excessive. We could change the size of flame and pot to heat only a small amount of water, conserving energy and making the process faster.

This is a simple example to demonstrate the principles of functional modeling of a technical process. Functional modeling of a technical process can be applied to any technical process or procedure that is performed in steps. Process modeling is an analysis tool that helps us recognize problems by releasing psychological inertia.

Chapter 7

Scientific Effects

Chapter Contents

Chapter Summary: *A key finding of the analysis of patents was that many innovative solutions used ideas from outside their own fields of science and engineering. The ability to search for scientific effects and phenomena outside our own field can therefore help us to innovatively solve problems. A scientific effect is a phenomenon that can be used to provide a desired function or action to solve a problem. To find alternative scientific effects, and maximize the number of ways we can find for performing the same basic function, we describe the function we are looking for in simple language, similar to functional language, we avoid high level technical terms. If available, search for alternative ways of performing a function or action using a scientific effect reference database, if not - search other sources of scientific phenomena.*

1. Introduction to Scientific Effects (also known as physical effects)

Classical TRIZ has four main solutions tools. These solutions tools provide concepts or ideas for solving a technical problem (note that in our main problem solving roadmap we also include a review of trends of evolution as solutions tool).

In the table below which shows solutions by problem model, we list the tools used to solve each model type. Scientific effects may be used to solve any of the four model types. For Su-field and technical contradictions we do not specifically list scientific effects as a model of the solution because our main solutions models are inventive standards and inventive principles.

Model of the Problem	Tool	Model of the Solution
Function Statement	Scientific Effects	Specific Scientific Effects
Su-field	Inventive Standards	Specific groups of Inventive Standards
Technical Contradiction	**Basic:** Contradiction Matrix and 40 Inventive Principles	Up to four specific Inventive Principles
	Advanced: ARIZ tools	ARIZ tools for solving a technical contradiction
Physical Contradiction	**Basic:** Separation, Satisfaction, Bypass	Satisfy, Bypass, Separate (supported by table of Specific Inventive Principles to Physical Contradictions and research of Scientific Effects)
	Advanced: ARIZ tools	ARIZ tools for solving a physical contradiction

Note that ARIZ is a classical solutions tool for solving technical and physical contradiction at the advanced level.

2. Scientific Effects

To solve technical problems and create innovative solutions, it is frequently necessary to be aware of and know how to apply various scientific effects. If you don't know the scientific effect, you can't solve the problem with it.

Problem solvers, engineers and scientists only know a limited number of scientific phenomena, and their knowledge is often confined to their own field of expertise.

A wider knowledge of physical, chemical, geometric, mathematical, biological effects and how to apply them would vastly increase problem solving capability. The analysis of patents revealed that many "inventions" were simply the result of implementing solutions using scientific effects from another field.

To close this knowledge gap, Altshuller initiated a database of scientific effects and phenomena to provide examples of scientific effects that problem solvers could reference and apply to problems. There are databases containing often more than 10,000 examples of scientific phenomena both in reference book and software form.

2.1 Examples of Scientific Effects

Scientific Effect or Phenomenon	Example of Application
Curie point – a magnet loses its magnetic properties at a specific temperature.	A valve controlled by a magnetic field can be actuated when the temperature reaches the Curie point.
Shape memory alloy	Spectacle frames are easily bent out of shape. Using shape memory alloy, they can be returned return to their original shape by heating.
Möbius strip is a band that is twisted to form one instead of two sides.	Using a Möbius strip for a conveyor belt makes it last longer because the entire surface get the same wear instead of one half of the surface.

Note that scientific effects are not the same as the 40 inventive principles or the 76 inventive standards. The 40 inventive principles are generic principles or ideas for solving technical contradictions. The inventive standards are ways of improving an interaction or "developing a system." Scientific effects are specific phenomena that are often used in one scientific field but not known in another.

2.1 Scientific Effects Databases

Scientific effects are simply phenomena that if known can solve our problem. If you have access to a "Scientific Effects Database," you can easily research scientific effects that help to provide solutions. Often they are cataloged by general function type, such as, how to "heat" something, how to "move" something. Or they are cataloged by generic function and object, such as how to "heat liquid substance," how to "heat solid substance", how to "move solid substance," how to "move gas," etc.

Similar to a standard reference dictionary, key action words can be used to search for scientific effects. For example, if you require a scientific effect that performs the function "dries," it is useful to look for the noun associated with the action word, such as "dryer" in addition to the verb "dry." It is a good idea to use a thesaurus to find the various closely related terminologies for the action you are researching.

If you don't have access to a scientific effects reference database then an alternative is to use any reference source such as internet search engines, scientific databases, patent databases, technology databases, etc.

Nature

It is often useful to try to look to nature for solutions (scientific effects). Nature has evolved solutions to specific problems over long periods, often producing the most effective solutions and designs. For example, the U.S. Navy initially used a weight that shifted between the bow and stern of the first submarines to direct the vessel up or down. This was ineffective because the weight would get stuck in a steep dive. The solution was discovered in nature, fish have moveable fins that direct them up and down. The concept was copied and moveable fins were installed on submarines.

Let's look at how to find scientific effects to solve a function statement problem and physical contradiction problem.

3. How to Search for Scientific Effects to Solve a Function Statement Problem

The problem is modeled, defined or formulated as "how to perform or improve a function" question. First we identify a function to be performed or improved using simple functional language.

Subject **- Function – Object**

The Function-Object part of the function statement is the part we focus on to make it as general as possible.

Let's say our problem is that we cannot dry clothes quickly in air. In functional language this is stated as:

Air – **removes – water** (insufficiently)

We should search for ways to remove water, or remove liquid, or remove one substance from another or separate substances.

Create a noun for a system that currently performs this function for example, drier, evaporator, desiccator, separator, moisture extractor.

If we are looking for the function to cool, we search for temperature reduction, a cooler, refrigerator, refrigeration, freeze, freezer, heat pump, conduction, conductor, etc.

If filter stops particles/dirt insufficiently, look for "filter" or "separates" to find alternative ways to separate. Identify synonyms for the basic general function that's being performed and try to find alternative ways of performing the same function. It may be useful to try specifying other scientific or engineering fields in the search. For example, "chemical drying", "biological drying," "radiation drying," etc.

128

Using the list of function-objects and nouns, search the available information sources.

Specific Problem Function Statement	Generic Function to Search	Alternatives Search Words
Water cools metal (insufficient).	How to: cool a substance.	Search: to cool, lose heat, heat exchanger, cooler, temperature, refrigerate, etc.
Sand erodes rock (harmful)	How to: eliminate erosion, stop removal.	Search: cover, protect, harden, shield, etc.
Belt moves boxes (insufficient)	How to: move an object.	Search: move, transport, etc.
Filter stops particles (insufficient)	How to: stop particles.	Search: filter separator, selector, etc. chemical filtration, physical, mechanical filtration, etc.

The objective is to find ideas for alternative ways of performing the same function but using scientific effects, outside of your scope of knowledge of field of expertise.

3.1 Algorithm for How to Search for Scientific Effects to Solve a Function Statement Problem

1. State the specific problem in the form of a Subject-Function-Object statement.
2. State the generic function to be searched for (examples shown in table above).
3. List alternative word search.

Specific Problem Function Statement	Generic Function to Search	Alternatives Search Words

4. Create ideas/solutions for performing the desired function in alternative ways.

3.2 Exercise: List five scientific effects that can be used to cool a can of soda.

4. How to Search for Scientific Effects to Solve a Physical Contradiction Problem

We chose to discuss scientific effects immediately after functional modeling because "scientific effects" is the main TRIZ tool for solving a function statement problem. Scientific effects are also a key tool used to solve physical contradictions (Chapter 9) as well as providing solutions to function statements. We recommend the user reads Chapter 9 in which physical contradictions are discussed before reading this section.

In Chapter 9 we list three ways to solve a physical contradiction, they are satisfaction, separation and bypass.

- To satisfy - a technical solution is found that solves the opposing requirements of the contradiction.
- To separate - opposing requirements (to be hot and cold, sharp and blunt, etc.) are separated in space, in time, or in relation to something or at system level.
- To bypass - a different way to solve the problem is found that does not require solving the conflict.

There are lists of specific inventive principles that suggest ways of separating in time, space, relation, satisfying and bypassing a problem (see Appendix 7: Table of Specific Inventive Principles to Solve Physical Contradictions). In addition to this list, we can also search for specific scientific effects that will provide ideas to solve the physical contradiction. Scientific effects that satisfy conflicting requirements, separate conflicting requirements, and bypass the problem.

4.1 Algorithm for How to Search for Scientific Effects to Solve a Physical Contradiction Problem

1. State the problem
2. State the physical contradiction
3. Describe the function to be researched (action-object and noun)
4. Document results of search for scientific effects
5. Solution

4.2 Examples of using Scientific Effects to Perform Satisfaction, Separation and Bypass to Solve a Physical Contradiction

Example 1 (Satisfaction)
1. **State the problem**: De-ionized high purity water is used throughout a manufacturing plant. Occasionally, there is bacterial growth in the water which impacts manufacturing. Because the water must remain pure, chemical additives which would normally be used to kill bacteria cannot be added to the water.
2. **State the physical contradiction**: Water must be chemically pure to be used for manufacturing and not chemically pure to kill bacteria.
3. **Describe the Function to be researched** (action-object and noun): Kill bacteria, exterminate, bactericide, sterilize, etc.
4. **Document results of search for scientific effects**: Ultraviolet (UV) light kills bacteria.

5. **Solution**: Pass the water through a section of clear pipe exposed to UV light to kill the bacteria, the light does not contaminate the water and bacteria will not grow – therefore the opposing requirements are satisfied.

Example 2 (Separation)
1. **State the problem**: The ends of two pipes must be connected by fitting one inside the other but the pipes are the same size and cannot fit one inside the other.
2. **State the physical contradiction**: The circumference of one pipe end must be large to allow the other pipe to fit inside and the circumference of the pipe must be small to firmly hold the inner pipe. The circumference of the pipe must be large and small.
3. **Describe the Function to be researched** (action-object and noun): How to fit pipes, connect pipes, join solid substances, expand solid substance, shrink, connectors, glue, attach, etc.
4. **Document results of search for scientific effects:** Thermal expansion and contraction of metal.
5. **Solution**: Thermal expansion will make the outer pipe large to allow the other pipe to be inserted; as it cools it will shrink and tightly hold the inner pipe.

    Separation of the opposing requirements in time, i.e. the pipe is large when being fitted, small after fitting.

Example 3 (Bypass)
1. **State the problem**: Electric cables in a house are used to connect computers to an internet connection. Installing wires involves much disruption and sometimes may not be possible.
2. **State the physical contradiction:** Wires must be present to provide communication path and not present to avoid physical damage.
3. **Describe the Function to be researched** (action-object and noun): connect a computer, electrical link, communication transmission etc.
4. **Document results of search for scientific effects**: wireless communication requires no physical connection.
5. **Solution**: Install Wi-Fi.

Chapter 8

Inventive Standards and Su-field Modeling

Chapter Contents

Chapter Summary: *A Su-field is similar to the function statement we discussed in Chapter 6. A function statement is a situation where A performs a function on B. A functional model of a technical system or technical process is normally made up of many components and interactions. A Su-field is the interaction between two objects via a "field" - Substance 2 (S2) acts on Substance 1 (S1) via a field, where S2 and S1 can be any "thing," an entire system, component, object, material, etc., but not a field.*

To create a Su-field model of a problem, a technical problem is "simplified" to be stated as "how to synthesize or improve the interaction between things S2 and S1 that interact by using field F." The problem is defined by the quality of the interaction similar to functional modeling; actions can be defined as useful, harmful, insufficient, excessive or absent.

This two component and field interaction (Su-field) when complete is also known as the "minimum working system." A problem is solved by first by transitioning the harmful, insufficient, excessive or harmful action into a useful effective interaction by referring to specific groups of the 76 inventive standards. When a useful effective Su-field interaction or "system" has been created using Classes 1 and 2 of the 76 inventive standard solutions, ideas for how to develop the "system" further and to become more ideal are provided by the additional Classes 3 and 5.

The 76 inventive standards are empirically derived ideas for how to solve problems based on knowledge from the patent database. They include information of the trends of evolution, the 40 inventive principles, scientific effects and common sense.

To use inventive standards, we will first state the problem as useful, harmful, excessive, insufficient or absent Su-field interaction. Then, using the Inventive Standards Flowchart (shown later in the chapter), consider the various inventive standard solutions you are directed to for each Su-field type, these are analogous solutions to your problem. The 76 inventive standards are separated into five solution groups. Class 1 and 2 provide initial solutions that form an effective complete Su-field by creating, destroying and developing Su-field interactions. Class 3 develops the solution, by providing ideas for how to "evolve the system" that are mainly based on learning from trends of evolution and Class 5 solutions make the system more ideal. Class 4 can be considered as a separate class of solution, devoted specifically to the improvement of measurement and detection.

Su-field modeling and inventive standards are also used to solve technical contradictions (Chapter 9) as well as improve an individual interaction. Altshuller's first method of solving technical contradictions was to use the 40 inventive principles and Contradiction Matrix (discussed in Chapter 9). He replaced the 40 inventive principles and Contradiction Matrix tool for solving technical contradictions with the System of Inventive Standard Solutions in the mid 1980's (the System of Inventive Standard Solutions is Altshuller's name for Su-field modeling and the 76 inventive standards).

We will use both the 40 inventive principles and Contradiction Matrix and the System of Inventive Standard Solutions as tools for solving a technical contradiction because they complement each other by providing more solutions than either does individually (this is discussed in detail in Chapter 9).

*We will treat the 40 inventive principles and Contradiction Matrix as a **basic** tool for technical contradiction problem solving and Su-field modeling and the System of Inventive Standard Solutions as an **advanced** tool for solving technical contradictions. In Chapter 9 it will be explained that we can express a technical contradiction as two technical conflicts in terms of opposing actions. Inventive standards solve technical contradictions by improving the harmful insufficient, excessive or absent action of a technical conflict.*

1. Su-field and Function Statement Models of an Interaction

Previously, in Chapter 6, we stated function and Su-field interactions are represented using different notation.

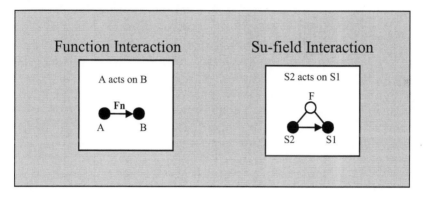

F = Field. Fn = Function.

There are two common ways to represent a Su-field: the "standard" and "barbell" notations.

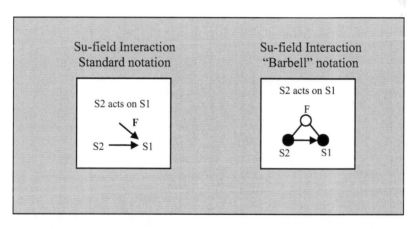

Both contain the same information, S2 acts on S1 via field F. We will use the "barbell" notation; it is closer to the notation we learned for functional modeling.

Note that when Su-field modeling the convention is, the subject is called S2 and the object, S1. S2 acts on S1 via field F. S2 and S1 are abbreviations for Substance 2 and Substance 1. In Su-field modeling the subject and object cannot be a field because fields are the means by which the action between two "things" takes place.

135

1. 1 Definition of a Substance and Field

We define a **substance** as any material "thing": animal, mineral, vegetable or combinations thereof. It can be in any physical state or form: plasma, gas, solid, liquid, powder, porous, paste, gel. It can possess any physical property: elastic, reflective, soft, sticky, conductive, reflective, ferro-magnetic, intelligent, etc.

When we look for inventive solutions we can consider a substance we "add" to create a solution to be "nothing." We can consider a "substance" to be: voids, hollows, cavities, clearances, gaps, space, capillaries, pores, holes, foam, etc.

The definition of **field** in TRIZ is a very loose. Because we are trying to free psychological inertia in a way that will enable us to search for alternative types of field, using the strict definitions like those that are used in physics and other sciences would limit our imagination.

The definition we use for a field is the means by which the action occurs. It includes all the strict definitions used for fields in science but we also include a wider type of definition. For examples, fields include a taste field, pressure field, smell (olfactory) field, stickiness field, biological, chemical, acoustical, optical field, etc.

MATCHEM is a simple mnemonic often used to remember some common field types shown in the table below.

Field Type	Examples
Mechanical	force (pulling, pushing, lifting lowering), shear, gravitational, tension, stress, strain, compression, friction, inertia, torque, centrifugal, centripetal. coriolis, buoyancy (Archimedes force), capillary, surface tension, elasticity, pressure, vacuum, hydraulic, aerodynamic
Acoustic	sound, vibration, ultrasound, shockwave
Thermal	thermal conduction, convection, radiation, thermal gradient, expansion, contraction
Chemical	oxidation, reduction, exothermic, combustion
Electric	electromotive force, electrostatic (attraction, repulsion),
Magnetic (and Electromagnetic)	magnetic, electromagnetic (light, microwave, x-ray, ultraviolet, visible, infrared, radio
Other Types	biological, atomic, molecular, sub-atomic forces, smell (olfactory), and any descriptive name for a field that aids imagination e.g. stickiness, taste, etc.

1.2 Definition of the Quality of an Interaction

In Chapter 6 where we describe functional modeling, we define the *quality* of an interaction by how effective we consider it to be. The convention commonly used to illustrate these actions by the specific connecting arrows is shown below. Useful, harmful, insufficient, excessive or absent are a qualitative assessment of the interaction, an opinion.

Functional Modeling

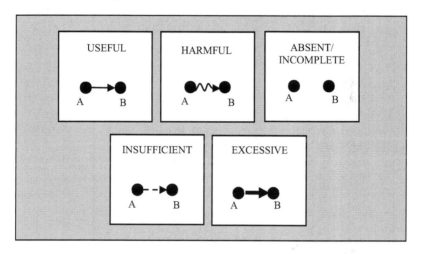

In Su-field modeling, Altshuller studied the patent database and classified five Su-field types. They are slightly different from those we used for functional modeling (ineffective is used instead of insufficient, no excessive is used and measure/detection is introduced). Note that absent is the same as incomplete.

Su-field Modeling

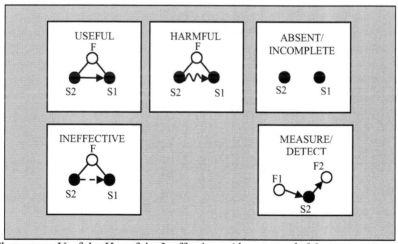

They are: Useful, Harmful, Ineffective, Absent, and Measurement and detection.

Instead of insufficient, ineffective is used. We will treat ineffective as being the same as insufficient.

Altshuller did not include an excessive interaction as a problem type when he created Su-fields and inventive standards. We will treat excessive in the same way as a harmful interaction (see Inventive Standards Flowchart later in this chapter).

Excessive interactions need to be reduced (partly removed) or partly neutralized. Harmful interactions need to be completely removed or neutralized. Many of the solutions for both are similar in concept.

We will define the quality of an interaction to match functional modeling with the inclusion of measurement/detection problems.

Altshuller's Su-field Quality of Interaction Descriptions	Our convention for Su-field Quality of Interaction Descriptions
Useful	Useful
Harmful	Harmful
Altshuller did not identify excessive, we will treat excessive as a harmful action when applying the inventive standard solutions	Excessive
Ineffective	Insufficient
Absent (or incomplete)	Absent
Measurement and Detection	Measurement and Detection

Note that the types of quality of interactions of Su-fields now match the quality of interaction of functional modeling with the exception of the definition of a measurement and detection problem. This makes using TRIZ simpler, we simply define the quality of an interaction in terms of functionality just as we defined them in functional modeling (with the measurement and detection separate classification).

Altshuller gave measurement and detection problems a class of their own in Su-field modeling. There is no measurement and detection class of problem in functional modeling. That does not mean all measurement and detection problems are to be formed as Su-field problems. They can be solved by creating function statements and searching for scientific effects that perform the function to provide a measurement or formulated into a contradiction.

Altshuller recognized there are a number of specific ways to help solve a measurement and detection problem when expressed as a Su-field problem and provided a list of 17 standard solutions.

1.3 Examples of Function Statements and Su-field Statements

Function: Hammer does not hit nail (absent/incomplete)

Su-field:
S2: Hammer
S1: Nail
A mechanical field F is needed to make an effective complete Su-field.

● ●
S2   S1

Function: Person smells flower - flower informs person (effective complete)

Su-field:
S2: Flower
S1: Person
F: Olfactory Field (smell)

F
S2   S1

Function: Hammer moves nail (insufficient)

Su-field:
S2: Hammer
S1: Nail
F: Mechanical Field

F
S2   S1

Function: Honey holds fly (excessive)

Su-field:
S2: Honey (like glue)
S1: Fly
F: Mechanical or stickiness field

F
S2   S1

Function: Compressed Air moves liquid (harmful)

Su-field:
S2: Compressed Air
S1: Liquid
F: Pressure Field

F
S2   S1

Function: Sugar informs tongue (effective complete)

Su-field:
S2: Sugar
S1: Tongue
F: Taste Field (or chemical)

F
S2   S1

2. How the System of Inventive Standard Solutions Works

Before we discuss inventive standards let's discuss the concept of how Su-field modeling and inventive standards work to provide us with innovative ideas. The relationship between Su-field models and inventive standards can be conceptually quite difficult and many struggle with it.

A specific technical problem with a known root cause is expressed or simplified to a simple problem interaction statement made up of two substances, "things" and a field. The problem interaction is unsatisfactory and identified as an absent/incomplete, insufficient, excessive, harmful or a measurement and detection interaction problem.

For example:

- If a pipe is leaking, we can express that problem as: pipe insufficiently stops water via a mechanical field.

The "76 inventive standards" is a list of generic solutions based on combinations of many of the 40 inventive principals, trends of evolution, scientific effects and common sense ideas based on the statistical analysis of the patent database. Those solutions are general models of solutions, many of which are ideas that by analogy can solve the problematic Su-field.

Su-field modeling operates in the same way we described previously for classical TRIZ tools.

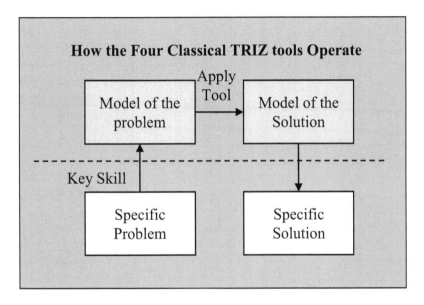

In the case of Su-field modeling, the Model of the Problem is a Su-field model (an absent/incomplete, insufficient, excessive, harmful or measurement and detection Su-field). The model of the solution is a list of specific sections of the 76 inventive standards which contains inventive standards we should consider as analogous solutions to the problem.

Later in this chapter, we will discuss a flowchart that we will use to direct us to the appropriate specific parts of the 76 inventive standards for each Su-field problem type. The 76 inventive standard solutions are organized into 5 classes and each class has a number of groups of individual inventive standard solutions. We will also discuss the organization of inventive standards later in this chapter.

The way we will use Su-field modeling and inventive standards is simple – first define the Su-field problem type (by the quality of the interaction or a measurement and detection problem) and second follow the flowchart to the generic solutions for the problem that are contained in the 76 inventive standards.

For example, if the problem were a harmful Su-field, the user is directed by the flowchart to first consider solutions in Class 1 Group 2 to create a useful effective Su-field, then to consider improving the solution by developing the "system" by using the solutions in Class 3, the user is also directed to solutions in Class 5 to consider how to make their solution more ideal.

Before we discuss the 76 inventive standards, how they are organized and the flowchart for directing us to the appropriate inventive standards to consider, let's discuss Su-field solution notation.

3. Su-field Solution Notation

In the detailed list of the 76 inventive standards (see Appendix 6), some of the solution descriptions are assisted in their explanation by a supportive Su-field model notation illustration of the solution. For example an inventive standard solution might say: use a double Su-field, or form a chain Su-field, or an internal or external complex Su-field or in the case of an incomplete Su-field problem, form a complete (effective) Su-field.

Note that only some of the 76 inventive standards solutions have a supporting Su-field model of the solution. Many solutions are too complex or are impossible to illustrate using the Su-field notation. Let's discuss this after reviewing the Su-field solution notation types.

3. 1 Complete Su-fields

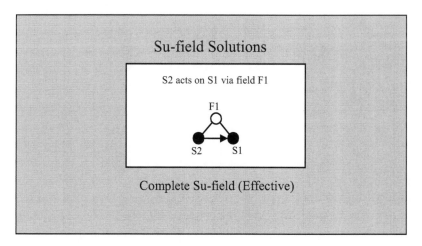

3.2 Double Su-fields and Chain Su-fields

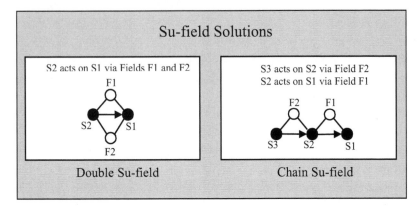

A double Su-field solution has two fields with the second field in parallel to the first. For example, a television set informs by light and sound in parallel.

A chain Su-field solution adds a third substance and a second field in series. For example, a person (S3) speaks (via audio field F2) into a megaphone (S2), the megaphone amplifies the voice; the stronger signal (F1) informs the crowd (S1).

3.3 External and Internal Complex Su-fields

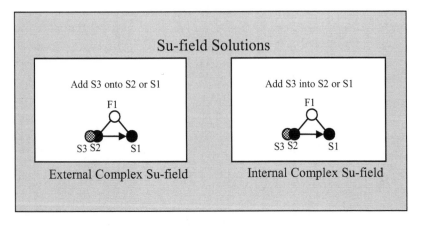

Adding a new substance **onto** S1 or S2 creates an External Complex Su-field. For example, putting insulation (S3) around pipe (S2) to stop heat loss.

Adding a new substance **into** S1 or S2 creates an Internal Complex Su-field. For example, adding salt (S3) into water (S2) to raise boiling point.

3.4 Measurement/Detection Su-field

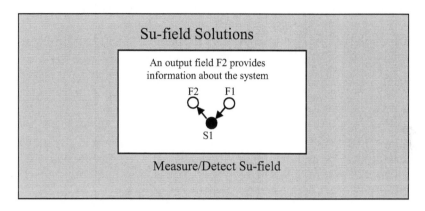

A measurement/detection Su-field is often represented as an output field F2 that provides information about substance S1 and/or input field F1.

3.5 When is Su-field Solution Notation Used?

The above sections, 3.1 to 3.4 describe the Su-field notation used in *some* of the 76 inventive standard solutions. They are used to help explain some of the solutions in terms of how the initial Su-field problem is transitioned into a solution.

This notation cannot be used for describing all of the solutions contained in the 76 inventive standards. It is mostly only helpful and used for Class 1 or Class 2 of the five classes of solutions (we will discuss the classes shortly).

The notation is not useful for many of the more evolved solutions. For example, inventive standard 2.2.4 is to dynamize (see Appendix 6). How can the solution "dynamize" be illustrated using the transformation of the triangle drawing of two substances and a field?

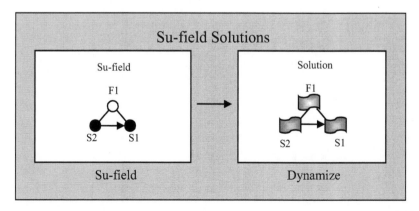

We could start making up our own notation and symbols, as shown in the diagram above. We created the symbol of a "waving flag" to indicate dynamization. In this solution both substances and the field are "dynamized." But this doesn't help us understand the solution any better than the description of the solution, which is to dynamize.

For inventive standard solution 3.2.1 - Transition to the micro level, we could draw the solution using smaller triangle or many small triangles or many S1's S2's or F's or even add to the notation, put an "M" next to each S2, S1 and F to indicate "micro." We could represent how to structurize substances (2.2.6) from 1 Dimensional to 2 Dimensional to 3 Dimensional by adding multiple D symbols or for inventive standard 3.1.1, how to represent the idea of Mono-bi-poly in a Su-field model?

The point here is that it does not increase the value of the 76 inventive standards as a tool to prompt ideas by us adding complex symbols to represent the 76 solutions, the majority of the solutions are best explained without a supporting Su-field diagram.

Su-field notation is simply a method for us to simplify our problem definition into a basic triangle subject-field-object interaction that is absent/incomplete, insufficient, excessive, harmful, or measurement/detection problem. The Su-field notation of *a few* of the solutions is assisted by explaining the solution in terms of substances and fields.

Let's summarize how the 76 inventive standards are organized before we discuss how to determine which standards to consider when creating innovative ideas to solve a problem.

4. 76 Inventive Standards Summary

The "76 Inventive Standard Solutions" of TRIZ are derived from the statistical analysis of the patent database. They are a list of inventive ideas derived from the trends of evolution, 40 inventive principles, scientific effects, and common sense ideas.

As discussed above, we define a Su-field problem as a harmful, insufficient, excessive and absent interaction or a measurement and detection problem. Solutions to Su-field problems are contained in the list of 76 Inventive Standard Solutions.

The Inventive Standard Solutions are organized into five classes. The classes are not organized by Su-field problem type; instead they are organized partly by Su-field problem type and partly by solution type.

4.1 Five Classes of the Inventive Standard Solutions

	Number of Solutions
1. Creation and destruction of a Su-field	13 standard solutions
2. Development of the Su-field model	23 standard solutions
3. Transition of the "system" to the supersystem and micro level	6 standard solutions
4. Standards for detection and measurement	17 standard solutions
5. Recommendations for more ideal application of the standards	17 standard solutions
Total:	76

In order to find the appropriate solutions for a harmful, incomplete, excessive, insufficient or measurement/detection Su-field problem we need a key to direct us to the appropriate standards to consider as solutions. We will use a flowchart to lead us to the appropriate sections of the 76 inventive standards for each type of problem.

Before we discuss the flowchart and learn how to find solutions by problem type, let's summarize the 76 inventive standards are organized. Note that this is a general summary of Appendix 6 which has details with examples of each the 76 inventive standards.

Class 1

Class 1.1 deals with the creation of a Su-field (addressing an **incomplete** Su-field) and the improvement of an **insufficient** Su-field.
- A substance or field needs to be added a Su-field to make it work or become effective.
    - o Complete the Su-field by introducing the missing substance or field or both.
    - o Introduce additives (substances/objects) to make Su-field work (temporary, internal, external/between or into the environment).
    - o Maximize/optimize the quantity of a substance or field that is used to make it more effective.
    - o Perform local intensification or weakening of the interaction to make it more effective.

Class 1.2 Deals with how to address a **harmful** Su-field.
- The Su-field interaction should be destroyed or neutralized.
    - o Introduce a new substance between S2 and S1 to block the harmful interaction (1.2.1).
    - o Introduce a substance that's modification of S1 or S2 to block or neutralize the harmful interaction (using a modified existing

145

substance 1.2.2 is a more ideal solution than introducing a new one).

- o Add a third sacrificial substance that will draw the harmful effect (1.2.3).
- o Introduce a second field to counteract the first (double Su-field) (1.2.4).
- o Turn off a magnetic field (1.2.5).

Class 2

Class 2 inventive standards strengthen and develop an **insufficient** interaction. In 2.1 we form a chain Su-field or a double Su-field by adding another substance to boost the weak field (like adding an engine). In 2.2 we try to improve the weak interaction by trying to evolve or change the existing resources (substances, fields, structure, timing) without adding more resources.

2.1 Deals with an insufficient **field.** The field is weak so move to a chain or double Su-field to "boost" the interaction.

- A chain Su-field is made by adding a field F2 and introducing a new substance S3 in series (next to S2 or S1). The new field enhances the existing one.
- A double Su-field is created by adding a second field in parallel (between S2 and S1) that compensates for the weak field. S3, the object that supplies F2 in parallel, by convention is not shown in the double Su-field model notation.

2.2 Also deals with an insufficient field, in this case by trying to develop the resources. Use the internal resources more effectively to improve functionality (without introducing new fields and substances by increasing the use of internal resources).

- Use a more evolved field (refer to the MATCHEM mnemonic).
- Transition to micro level.
- Use porous or capillary materials.
- Dynamize (joint-chain-elastic-liquid-gas-field).
- Increase the coordination of rhythm of substances and fields.

2.3 Tries to improve an insufficient interaction by redistributing energy more effectively.

- Use frequency: continuous actions, pulse, constant frequency, resonance.
- Structure fields and substances.
- Coordinate frequency, rhythm, timing of fields.

2.4 Ferro magnetic solutions are used to try to improve an insufficient interaction.

Class 3

In Class 1 and 2 we improved the interaction by adding to or modifying resources. Now we look for opportunities for improving functionality by developing the system. We look to the supersystem and micro levels.

3.1 Function moves to a supersystem
- Mono-bi-poly: simplest transition to the supersystem, increase the functionality of the interaction by combining two or many *similar* substances.
- Evolve links: improve functionality by combining substances together.
- Functions become more diverse: improve functionality by linking different substances e.g. pencil with eraser.
- Trimming: improve functionality by having substance perform additional actions and removing the need for one or more substances (components).

3.2 Transition to micro level
Develop the system by transitioning its operation to the grain, particle, molecule, subatomic, etc., level.

Class 4

Class 4 deals with how to improve or establish methods of **measurement and detection**.

4.1 Change the system so there is no need to measure (use workarounds). Measure copies, marks or labels to make measurement easier (e.g. barcode labels and scanner). If you cannot measure continuously, make discrete measurements.

4.2 Measure indirectly or measure additives or derivatives.

4.3 Use resonance measurements.

4.4 Use ferro-magnetic fields (Curie point etc.).

4.5 Transition to bi and poly systems (measure two or more) or use first and second derivatives (measure speed and acceleration instead of distance).

Class 5

Class 5 deals with how to best apply the inventive standards in Classes 1 through 4. The most ideal way (minimum cost maximum functionality) is suggested.

- To introduce a substance:

    o Use workarounds/indirect ways
        - Add "nothing," a void or hole instead of a substance
        - Add a field instead
        - Add external instead of internal additive
        - Use small concentrated amounts
        - Use temporary instead of permanent additives
        - Use a copy or model
        - Use something from the environment or something that disappears
    o Use many small additives instead of one big one
    o Use inflatables

- Use fields (consider using fields from the environment)

- Make use of the phases of substances, (gas instead of liquid, solid instead of gas, etc.)

- Use substances that control themselves

- Use amplification

- Use higher or lower forms of substances (e.g. electrolysis to release hydrogen from water)

Altshuller includes many ferromagnetic solutions (used heavily in Sections 2.4 and 4.4).

Ferromagnetism is an interesting area of scientific effects because it provides a way of introducing fields and substances together as one, but in terms of finding practical solutions it may be considered to be over emphasized.

5. Which Inventive Standards to Choose

The Inventive Standards Flowchart directs us to specific groups of solutions depending on the type of Su-field problem (absent, harmful, insufficient, excessive, measurement/detection).

The flowchart below is a starting point. The inventive standards are *statistically* derived, if a good solution is not prompted by the inventive standards suggested by the flowchart, then other standards may be considered to try to prompt ideas.

5.1 Inventive Standards Flowchart

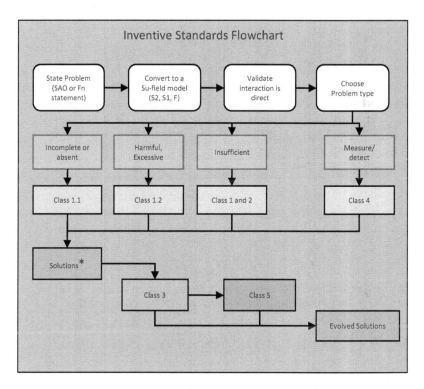

* Solutions are created by using Class 1.1, 1.2, 1, 2 and 4 to form an effective complete Su-field or an effective measurement and detection Su-field. The effective Su-field or "system" may be further developed by Class 3 and made more ideal by Class 5 inventive standards resulting in more evolved solutions.

In our flowchart, "excessive" is treated as a "harmful action." The suggested solutions that block or neutralize a harmful action are used to prompt ideas for how to partially block or neutralize an excessive action.

In the flowchart we start by forming the problem as a subject-action-object statement (a function statement could be used from any interaction in a functional model of a technical system or technical process); this is generally found to be the simplest and most intuitive construction for a Su-field.

After the Su-field problem is formed, we perform a check that the interaction between S2 and S1 in the Su-field is direct before considering the inventive standards that we are directed to by the flowchart.

5.2 Validating the interaction between S2 and S1 is direct when defining a Su-field problem

When the Su-field problem is formed the interaction should be direct, meaning the problem field should be directly between the subject and object. To check the action is direct, create a functional model of the Su-field that is formed by the subject-action-object or function statement. Check whether there are any substances and fields that should be inserted between the S2 and S1. If there are, the Su-field is not correctly defined and solutions will be ineffective. The model should be re-evaluated to define the Su-fields correctly.

An example of not defining the Su-field directly is: a golfer (person) must hit his ball over some trees but he cannot hit it high enough. We define the Su-field interaction initially as:

- Golfer insufficiently moves ball or golfer moves ball via a mechanical field insufficiently.

We should validate the interaction is direct by building a functional model between the subject and object to check if the interaction is direct. By building a functional model from subject to object, in this case we find that the interaction is not direct and therefore not correctly defined.

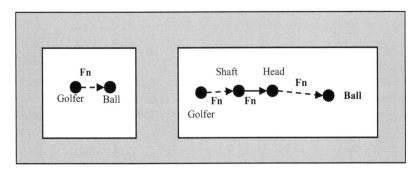

We find that the golfer moves shaft (of club), shaft holds head (of club) and head moves ball. We could define "golfer moves shaft" as insufficient or "head moves ball" as insufficient. These two new problems are the interactions we can choose to work on to find a solution. Making use of functional models is a key aid to accurately defining and validating Su-field problem models.

6. Inventive Standards Solve Problems but also Evolve the System

It's useful to think of applying the inventive standards in three steps:

1. Solutions for how to improve a problem Su-field interaction.
2. How to develop or evolve the interacting Su-field system using Class 3.
3. How to make the Su-field solution more Ideal using Class 5.

Problems are initially stated as negative interactions (incomplete, excessive, insufficient, harmful) which are solved by Classes 1 and 2 or measurement and detection problems (which are solved by Class 4) and then suggestions as to how to improve the solution by evolving, transitioning or developing the "system" are given in Classes 3 (where "system" means the solutions created by applying Class 1, 2 or 4). Class 5 provides ideas for how to make the solution more ideal.

If there is no problem, no incomplete, excessive, insufficient, harmful Su-field then an effective complete Su-field can be developed and made more ideal by Class 3 and 5 solutions. In other words we can improve technical systems by thinking of them in terms of effective complete Su-fields and by applying Class 3 and Class 5 solutions we can develop them and make them more ideal.

So, inventive standards provide ideas not only for solving a specific interaction problem but also for improving interactions by considering them as a system.

In Chapter 9 we will learn how to convert a technical contradiction into two technical conflicts. Each conflict is made up of a positive and negative action. Su-field modeling and inventive standards can be used to identify solutions to improve the negative actions and hence used to solve contradictions.

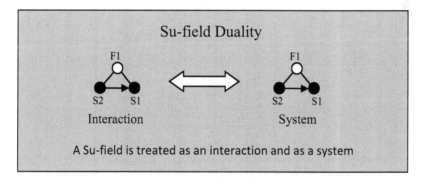

Let's look at basic examples before we discuss the algorithm for applying inventive standard solutions. Note that this algorithm will also apply to improving the negative action of a technical conflict.

6.1 Examples

Example 1
A garden hose takes too long to water a lawn, its stream is narrow and there is a large field to cover. The Su-field problem is "hose insufficiently directs water via a mechanical field."

Using the flowchart, for an insufficient interaction, we are guided to Class 1 or Class 2 initially. We consider each idea in turn (see Appendix 6 for detailed list of all inventive standards).

- From Class 1: we could introduce an additive between the water and hose. For example, a nozzle might improve the spread.
- From Class 2: we could dynamize, changing its spray direction in time; we coordinate rhythm by spraying only at night to stop evaporation of the water.

These are initial solutions, Class 3 and 5 develop the solutions further, make them more ideal.

- From Class 3: we could evolve the system using the trend of mono-bi-poly. Use many nozzles by creating a sprinkler system.
- From Class 5: we could try to use gravity to spread water more, dig an irrigation trench, use the wind to spread a finer spray etc.

Example 2
The hose pipe is leaking, (harmful)

The classification of the quality of interaction is open to interpretation. We could see the problem as "pipe insufficiently holds water" or "pipe harmfully passes water." Let's consider the interaction as harmful. We are initially directed to Class 1.2

- From Class 1.2: we get the idea to place a patch (S3) between the water and hose.
- From Class 5: our idea is to freeze the water in the region of the leak (more ideal).

7. Algorithm for Applying System of Inventive Standard Solutions

The recommended algorithm is to create a problem subject-action-object statement first. Use a functional model or create a function statement problem by making a functions statement (for example pipe insufficiently holds water, etc). Then check the interaction is direct. Once you have made sure the interaction is direct use the Standard Solutions Flowchart to progressively consider the suggested solutions to create ideas.

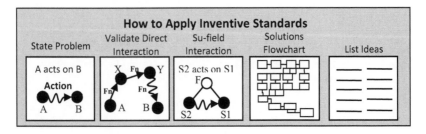

152

Step 1 State problem as Subject-Action-Object statement and assess the quality of the interaction.
- A does not act on B (incomplete or absent)
- A insufficiently acts on B
- A excessively acts on B
- A harmfully acts on B
- Information is not or is insufficiently measured or detected.

Step 2 Validate that the interaction is direct between A and B.
Check that there are no intermediate substances or interactions that link between S2 and S1.

The Su-field must define the *direct* interaction between the S2 and S1. Often, due to psychological inertia we inaccurately define the problem. Making use of functional models is a key aid to accurately defining Su-field models.

Step 3 Identify S2, S1, the field and re-state the type of interaction using the format
- A function B (quality of interaction) via field F1

For example,
- Paper insufficiently holds water via a mechanical field (news paper doesn't hold water well).

Step 4 Follow the Inventive Standards Flowchart.
The flowchart is a guideline. For additional ideas use the rest of the inventive standards to prompt ideas.

Step 5 List ideas prompted by the inventive standards.
Add them to your "solutions bank."

7.1 Examples of Applying System of Inventive Standard Solutions

Example 1
Problem A golfer (person) must hit his ball over some trees but he cannot hit it high enough. Create ideas to increase the height the ball travels.

Step 1 State problem as subject-action-object statement and assess the quality of interaction
We could define the problem as:
- golfer insufficiently moves ball (golfer moves ball via a mechanical field insufficiently).

Step 2 Validate the interaction is direct between S2 and S1.
We check if the interaction is direct by building a functional model between the subject and object.

153

The Su-field must define the direct interaction between the Subject and Object. The field must be direct. Often, due to psychological inertia we inaccurately define the problem.

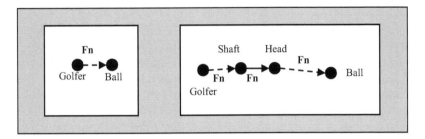

We find that the golfer moves the shaft (of the club), the shaft holds the head (of the club) and the head moves the ball.

By building a functional model of the interactions between the subject and object we more accurately define the Su-field problem. In this case we defined two interactions that we could improve.

- The person insufficiently directs the shaft via a mechanical field.
- The head of the club insufficiently moves the ball via a mechanical field.

In this example let's choose to pursue the interaction "head of the club insufficiently moves ball via a mechanical field."

Step 3 Identify S2, S2, the field and re-state the type of interaction using the format

- A function B (quality of interaction) via field F1

head moves ball (insufficiently)

S2 Head
S1 Ball
F: Mechanical Field (force)

Head of club insufficiently moves ball via a mechanical field.

154

Step 4 Follow the Inventive Standards Flowchart.

For an insufficient field we are directed to Class 1 or 2 then Class 3 and then Class 5.

Step 5 List ideas prompted by the Inventive Standards.
Add them to your "solutions bank."

1.1.3 Introduce additives (S3) temporarily or permanently into or onto or between S1 and S2. Our idea is to place a wedge that changes the angle of the head to gain lift.

2.2.4 Dynamize. Our idea is make the shaft more flexible or jointed to increase the speed of the shaft at impact.

3.1 Transition to the supersystem. Our idea is to move to a bi or poly system. Make a set of many interchangeable wedge heads for one club (easier to carry than many clubs).

5.2.2 Use a field from the environment. Strike the ball when there is wind to move it further.

Example 2
Problem A lead pipe needs to be bent but it collapses instead of bending.

Step 1 State problem as a subject-action-object statement and assess the quality of the interaction

Something (S2) bends pipe (S1) harmfully.

Step 2 Validate the interaction is direct between S2 and S1.
Ensure that there are no intermediate objects or functions between S2 and S1.

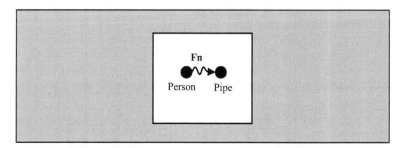

Person bends pipe. In this case the interaction is direct, there are no hidden links.

Step 3 Identify S2, S1, the field and re-state the type of interaction using the format

- A function B (quality of interaction) via field F1

person bends pipe (harmfully)

S2 Person
S1 Pipe
F: Mechanical Field (force)

Person bends pipe harmfully via a mechanical field.

Step 4 Follow the flowchart.

For a harmful field consider Class 1.2 of the inventive standards then Class 3 and then Class 5.

Step 5 List ideas prompted by the Inventive Standards

1.2.3 Introduce a "sacrificial" substance to draw off the harmful action.
Prompted by the inventive standard we create the idea to insert "bendable" metal pipe inside the pipe. The pipe will bend without collapsing, the metal added can later be removed (e.g. by chemical etching). This is similar to adding an internal additive into S1.

3.1.1 and 3.1.2 instead of bendable metal use a spring inside the pipe (many flexible parts linked together).

Example 3
Problem It is necessary to improve the drama of a movie by showing a bullet come directly from the barrel of a gun towards the camera. How is it possible to film (record/measure) the bullet without the camera being destroyed?

Step 1 State problem as subject-action-object statement and assess the quality of interaction.

Camera stops bullet (harmfully) via a mechanical field

Step 2 Validate that the interaction is direct between S2 and S1

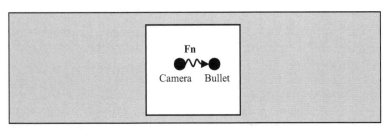

Camera stops bullet (harmfully) via a mechanical field. By building the functional model, we also reveal that there is another interaction present. Bullet informs camera via a light field this is a useful field.

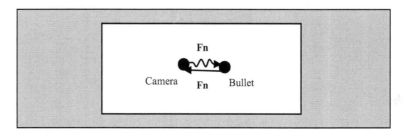

Bullet informs camera (via a light field).
There are no "hidden links."

Step 3 Identify S2, S1, the field and re-state the type of interaction using the format
- A function B (quality of interaction) via field F1

Camera stops bullet (harmfully)

S2 Camera
S1 Bullet
F: Mechanical Field (force)

Camera stops bullet via a mechanical field.

Step 4 Follow the flowchart.
For a harmful field consider Class 1.2 of the inventive standards then Class 3 and then Class 5.

Step 5 list ideas prompted by the Inventive Standards.
1.2.3 Introduce a sacrificial substance (S3) to absorb or "draw off" the effect of the harmful action. In this case our idea is to place a mirror (S3) in the path of the bullet and film the image in the mirror.

Note that we could also address this problem as a measurement and detection problem, we are measuring (recording information) about the bullet. We therefore can consider the solutions in Class 4. Inventive standard 4.1.2 directs us to measure a copy or an image. Again we get the idea to use a mirror to reflect the image of the bullet. The bullet will destroy the mirror and not the camera.

For a list of all 76 inventive standards with examples - see Appendix 6.

Chapter 9

Contradictions and ARIZ Tools

Part 1

Basic Contradiction Problem Solving

Chapter Contents

Introduction and Basic Contradiction Problem Solving
1. Contradictions Introduced
    1.1 Contradictions in Relation to the Five Levels of Invention
    1.2 Three levels of Contradiction
        o   Administrative
        o   Technical Contradiction
        o   Physical Contradiction
    1.3 Technical Contradiction- Two Technical Conflicts
    1.4 Why we form Technical Contradictions
    1.5 What if the problem does not form a contradiction?
    1.6 Physical Contradiction - One Physical Conflict
    1.7 Why we form Physical Contradictions
2. Solving Contradictions Overview
    2.1 Basic Contradiction Problem Solving
    2.2 Advanced Contradiction Problem Solving
    2.3 ARIZ Overview
    2.4 Overview of Contradiction Problem Solving Flowchart
    2.5 Contradiction Problem Solving Flowchart – Simplified ARIZ
3. Technical Contradictions Basic
    3.1 The Contradiction Matrix and the 40 Inventive Principles
        o   The 39 Matrix Parameters
        o   The 40 Inventive Principles
        o   Contradiction Matrix
    3.2 Basic Contradiction Diagram
    3.3 Algorithm for Application Inventive Principles and Contradiction Matrix
        o   Examples of Solving Technical Contradictions (using Basic Tools)
4. Physical Contradictions Basic
    4.1 Some general examples of Physical Contradictions:
    4.2 Forming Physical Contradictions (Antonym Pairs)
    4.3 Preliminary Physical Contradiction Formation
    4.4 Why do we form a Physical Contradiction?
    4.5 How to Solve a Physical Contradiction
        o   Satisfaction
        o   Bypass
        o   Separation Principles

Chapter Summary: *From the study of patents, researchers concluded that a breakthrough inventive solution (Level 2 and up) is the result of solving a contradiction. To achieve a breakthrough solution, we should define and solve a contradiction.*

Altshuller identified three levels of contradiction: an administrative contradiction, a technical contradiction and a physical contradiction. Administrative contradictions do not help solve problems. A technical contradiction expresses a problem in the form of a dilemma, typically by forming an "if-then-but" statement - if I do this, then "something good" happens but "something bad" happens as a result. For example, if the aircraft wing is large then lift increases but weight also increases. A physical contradiction expresses a problem in the form of an opposing physical requirement, something must be "A" for the good thing to happen but it must be "anti-A" to eliminate the bad thing. So "something" must be A and anti-A, where "A" and "anti-A" are opposite physical requirements. They form antonym pairs for example, hot and cold, large and small, present and absent. Using the example of the aircraft wing, we can create a physical contradiction - the aircraft wing must be large to increase lift but small to reduce weight. The wing must be large and small. It is by solving these technical and physical contradiction problems that we create breakthrough solutions.

An administrative, technical and physical contradiction are different ways of expressing the same technical problem, typically we do not state the administrative formulation. To systematically create a breakthrough inventive solution, a problem is stated initially as a technical contradiction then as a physical contradiction. A technical contradiction is actually made up of two

160

technical conflicts. A physical contradiction is made up of only one physical conflict.

We will define two levels of tools for solving contradictions - basic and advanced.

At a basic level, we apply the first major tool devised by Altshuller for solving technical contradictions, the Contradiction Matrix. In the Contradiction Matrix a technical contradiction is defined in terms of 39 general opposing parameters and the solutions defined by 40 general inventive principles. At the basic level, we form physical contradictions by identifying antonym pairs and we solve the physical contradiction by methods of separation, satisfaction or bypass.

A key new analytical tool, the "Contradiction Diagram" is introduced for analyzing contradictions, it helps us and state the technical contradiction as opposing parameters and formulate the technical contradiction in terms of opposing actions which is needed for applying the advanced tools for solving both technical and physical contradictions.

Basic level tools will provide breakthrough ideas but the advanced tools generally provide a greater number of more specific ideas by systematically considering all available resources for a solution. Many users of TRIZ only ever apply the basic tools for creating and solving technical and physical contradictions described in Part 1 of this chapter.

In Part 2 of this chapter we discuss the advanced level tools for solving technical and physical contradictions. The advanced level uses the key tools of ARIZ (the Algorithm for Inventive Problem Solving). ARIZ is the core algorithm of TRIZ. It is highly complex and contains many steps. ARIZ is discussed in Appendix 8.9, where we provide an example and template for ARIZ-85C the last version of the algorithm approved by Altshuller. An alternative to using ARIZ is to simply apply the basic tools discussed here in Part 1 to solve an inventive problem, and if needed, apply the advanced tools listed in Part 2 of this Chapter. These tools are put together to form the "Contradiction Problem Solving Flowchart – Simplified ARIZ" see Section 2.4 and 2.5 of Part 1 of this Chapter and the "Simplified ARIZ the Algorithm for Solving Inventive Problems," an 18 step process that describes each step of the flowchart, see Part 2 Section 7 of this chapter and Appendix 8.8.

Introduction and Basic Contradiction Problem Solving

1. Contradictions Introduced

In Chapter 1 we discussed the main conclusions from the study of millions of patents that led to the creation of TRIZ tools:

161

- Solutions could be classified into five levels. The "Level of Inventiveness" or "degree of difficulty" increasing from 1 through 5.
- Inventive solutions, Level 2 and up, contained the solution to a contradiction.
- The same few inventive principles were used to solve technical contradictions.
- Problems in one field had already been solved in another.
- Trends of technical evolution - how features of technical systems evolve and progress over time, were highly predictable.

It was from these findings that Altshuller and his colleagues developed the main classical TRIZ **solutions** tools. When a specific problem has been identified (normally after applying analytical tools) there are four main classical TRIZ solutions tools we can use to create ideas (we can also search trends of evolution).

Classical Solutions Tools:

Model of the Problem	Tool	Model of the solution
Function Statement	Scientific Effects	Specific Scientific Effects
Su-field	Inventive Standards	Specific groups of Inventive Standards
Technical Contradiction	**Basic:** Contradiction Matrix and 40 Inventive Principles	Up to four specific Inventive Principles
	Advanced: ARIZ tools	ARIZ tools for solving a technical contradiction
Physical Contradiction	**Basic:** Separation, Satisfaction, Bypass	Satisfy, Bypass, Separate (supported by table of Specific Inventive Principles to Physical Contradictions and research of Scientific Effects)
	Advanced: ARIZ tools	ARIZ tools for solving a physical contradiction

Note that we list ARIZ as a classical solutions tool for solving technical and physical contradictions at the advanced level.

In preceding chapters we discussed ways to formulate a specific problem as a function statement or Su-field. This allows us to search for and consider specific scientific effects and inventive standards that may solve our problem. In Part 1 of this chapter we will learn the basic tools of how to form and solve problem as a technical and physical contradiction. In Part 2 of this chapter we will discuss the advanced tools used in the ARIZ-85C algorithm for forming and solving technical and physical contradictions.

1.1 Contradictions in Relation to the Five Levels of Inventiveness

Altshuller classified the patents into five Levels of Inventiveness ranging from basic routine improvements to highly innovative patents that require new scientific discovery.

Level 1 and many of the Level 2 solutions did not involve solving a contradiction.

Level 1
Obvious or Routine Solution. The solution method is well known within its field (industry, technology) and often involves a simple adjustment or basic optimization. TRIZ does not class Level 1 as inventive, since it does not require breakthrough thinking; Level 1 is an improvement that does not solve a contradiction.

Example
- If I use very narrow skis then I sink in the snow. Solution: use wider skis. This is a Level 1 solution. There is no opposition to an obvious improvement solution. There is no contradiction.

A contradiction would be formed if there was opposition to widening. A contradiction is formed when something opposes an improvement. For example:

- **If** I widen my skis **then** I don't sink in the snow **but** I move slower with wide skis.

The improvement made by widening the skis is opposed by slowing the speed of the skis. Many Level 2 solutions require solving a contradiction. Level 3-5 solutions by definition solve a contradiction. Let's look again at our definition of Level 2-5 solutions.

Level 2
Level 2 solutions *may* involve solving a contradiction but it is not required. The solution is not well known within the industry or technology. It doesn't use knowledge from other industry or technology but requires creative thinking to provide the solution. Level 2 solutions may involve solving a contradiction but not a significant one.

Level 3
Significant improvements are made to an existing system. Normally this requires *engineering knowledge* from other industries or technologies. The solution resolves a contradiction using knowledge from other industries or technology.

Level 4

Solutions use *science* that is new to that industry or technology. The move to a new scientific effect or phenomenon eliminates the contradiction and usually involves a radical new principle of operation.

Level 5

Solutions involve discoveries of new scientific phenomena or new scientific discovery. Contradictions are solved by discovering and applying new scientific phenomena. Level 5 solutions can lead to the creation of many new products, industries or technologies which in turn creates many new inventions at Levels 4, 3, 2 and 1 (e.g. discovery of solid state transistors).

Not all Level 2 solutions solve a contradiction. We can think on contradictions starting somewhere between Level 1 and Level 2, at around level 1.5. For convenience we will say contradictions are solutions Level 2 and up.

1.2 Three levels of Contradiction

Above we mentioned one of the three types of contradiction defined in TRIZ. These are three levels in which a contradiction may be stated. They are (in hierarchical order):

- Administrative Contradiction
- Technical Contradiction
- Physical Contradiction

Administrative is considered the least well defined, it is far from being a specific problem; a physical contradiction is considered a very specifically defined contradiction.

An administrative contradiction is a simple conflict statement that can be expressed in the general form: I have a goal but don't know how to achieve it. The contradiction arises from the dilemma - I want to do something but I don't know how to do it. Examples of administrative contradiction statements are:

- I want to increase the speed, but I don't know how to.
- I want to improve weight, but I don't know how to.
- Heat loss from the surface should be improved, but I don't know how to reduce it.

An administrative contradiction is fairly meaningless in terms of defining a practical problem to solve, and is not used for creating solutions. It needs to be evaluated and re-stated as a specific technical or physical contradiction problem statement that can be solved.

A technical contradiction is formed when there is an obstacle to making an improvement. We try to make an improvement so that "something good" happens but "something bad" that we don't want, happens as well. A technical contradiction is usually created in the form of an **"if-then-but"** statement. It has three parts:

- If (state what action or change could be implemented to create the desired effect)
- Then (state the positive or desired effect)
- But (state the negative or undesired effect)

The "if-then-but" statement expresses a technical contradiction.

o **If** I do this, **then** something good happens **but** something bad happens too.
o **If** I implement a change, **then** a positive effect occurs **but** a negative effect also occurs.

Administrative contradictions are seen as hierarchically above technical contradictions in that, to solve the administrative contradiction we would formulate it into a technical contradiction or physical contradiction.

Similarly a technical contradiction is often considered hierarchically above a physical contradiction because a physical contradiction can be derived from a technical contradiction by expressing the conflict as an inherent need for a resource to possess two opposite physical characteristics to satisfy the conflicting requirements of the technical contradiction - a resource or part of a resource should be hot and cold, big and small, etc. (we shall discuss how to convert a technical contradiction to a physical contradiction in detail below).

A physical contradiction is formed when a system, component, part of a component, etc., has opposite physical requirements. For example, the handle must be rigid and flexible, the cup must be hot and cold, the blade must be sharp and blunt. Rigid and flexible, hot and cold, sharp and blunt are each opposite requirements of a single parameter (rigidity, temperature and sharpness), they form antonym pairs. Two mutually opposite physical states are required to solve the problem, hence the name physical contradiction. One or more physical contradictions can be created from a technical contradiction statement. We will learn how to do this below.

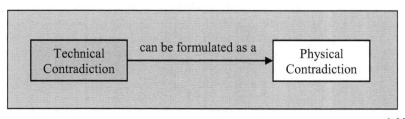

165

1.3 Technical Contradiction – Two Technical Conflicts

It is important to understand that there are two problem situations created by an "if-then-but" statement.

o **If** I make a change, **then** a good action occurs **but** a bad action occurs too.
o **If** I don't make the change, **then** the bad action doesn't happen, **but** the good thing doesn't happen either."

A technical contradiction is actually made up of two problem situations. We will refer to each problem situation as being State 1 and State 2. The problem in each state is the mirror image of the problem in the opposite state. Rather than name the problem in each state both as a contradiction, we will call them technical *conflicts* TC-1 and TC-2.

A technical contradiction is made up of two opposite technical conflicts, TC-1 and TC-2. *We will define a conflict as one half of a contradiction* where TC-1 and TC-2 are the mirror image of each other in opposite states (State 1 and State 2). For example:

o TC-1: **If** power is high **then** the fan speed is high, **but** the fan is noisy.
o TC-2: **If** power is low **then** the fan is not noisy, **but** the fan speed is slow.

The two states are State 1 (high power) and State 2 (low power). TC-1 and TC-2 are the "mirror image" of each other. Let's try another example (a hammer and nail):

State 1 is high force
o TC-1: **If** force applied by the hammer is high **then** the insertion speed is high **but** the bending rate of nails increases.

State 2 is low force
o TC-2: **If** force applied is low **then** there is no bending **but** insertion speed is low.
All technical contradictions can be formed into two technical ***conflicts.***

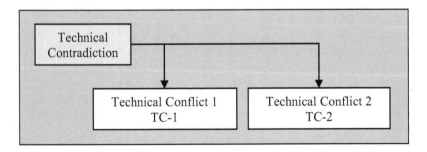

Try converting these "if-then-but" technical contradiction statements into two technical conflicts by describing the opposite state.

TC-1
1. **If** the wood is denser, **then** the strength of the chair increases, **but** the chair gets heavier.
2. **If** the chemical is more acidic, **then** the particles are removed better, **but** the surface gets damaged.
3. **If** power of the plasma is increased, **then** the directionality of the ions is more vertical, **but** more of the photoresist mask is eroded.

TC-2 for each of the examples is:
1. **If** the wood is less dense, **then** chair is lighter, **but** the chair is weak.
2. **If** the chemical is less acidic, **then** the surface does not get damaged, **but** particles are not effectively removed.
3. **If** power of the plasma is reduced, **then** less of the photoresist mask is eroded, **but** the directionality of the ions is less vertical.

1.4 Why we form Technical Contradictions

When we define our problem as a contradiction, we do not only capture the goal of a problem but also define what is opposing or blocking it. Our solution to a contradiction must provide the good effect and eliminate the bad effect, which requires breakthrough thinking. Often (if not using TRIZ) we don't clearly recognize what opposes the good effect or don't see that the good effect is being opposed.

Creating the contradiction sets us on the path to creative innovation. Instead of thinking of a compromise solution we are directed to solve the problem with no compromise. We are directed to a higher level of solution than Levels 1 or 2 and target a more innovative breakthrough solution since breakthrough inventive solutions solve contradictions.

For example, if we do not form a contradiction:

- Problem: How to increase the number of passengers on a bus?
- Solution: Put more chairs in, use smaller chairs, make the bus a bit longer, add a second level, ask passengers to stand. These are compromise solutions. They don't need higher level breakthrough thinking.

Our ideas tend to be those of compromise and are restricted by psychological inertia (we subconsciously assume the bus can be made only a bit longer).

Now let's form a technical contradiction (which is two technical conflict statements TC-1 and TC-2)

- TC-1: **If** I increase the length of the bus **then** the bus can carry more passengers **but** the bus cannot easily turn corners.

- TC-2: **If** I decrease the length of the bus **then** it can easily turn corners **but** it will carry less passengers.

Now we can try to solve the conflict in both cases.

- TC-1: How to make a **long bus** turn corners? One solution is to make it flexible, make a joint in the middle. This is a breakthrough solution.
- TC-2: How to make a **short bus** carry more passengers? Make it taller, make it wider but only on the second level (asymmetric in shape). Make the top deck dynamic so it can rotate avoid other buses as it turns.

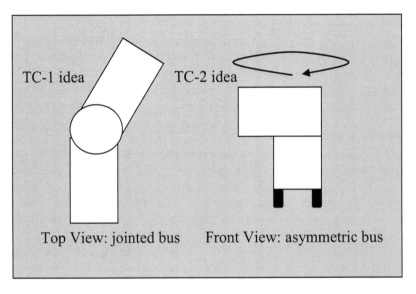

By forming a contradiction we identified a negative effect, something that blocks or opposes the positive improvement. This leads to more innovative breakthrough thinking. Instead of looking for compromise and finding compromise solutions, the creative thinker actively seeks out contradictions to create breakthrough solutions. Shortly we will discuss specific tools defined to help us create ideas for how to solve contradiction problems.

Notice we get two sets of ideas, those associated with TC-1 state and those associated with TC-2 state. Normally when we define the two states for a technical conflict, one state better supports the **main useful function** of the system. In this case the main useful function of a bus is to transport passengers. This is better achieved in State 1 where the bus is longer as opposed to State 2 where the bus is shorter.

Typically the solutions derived from the state in which the main useful function is better delivered are less radical than those for the opposite state.

168

Normally the less radical technical conflict is pursued for ideas first then the more radical technical conflict that least best supports the main useful function is pursued. This is because the less radical solutions usually provide solutions that require less change to the system, the introduction of fewer new components, etc. More ideal solutions are created by focusing on solutions that require minimal changes to the system. In our example an asymmetric bus is highly radical, leading in this case to an idea that is somewhat impractical. This, however, is the point, often we are led to solutions that are very radical and are very innovative, outside of our psychologically inert thinking.

By creating "if-then-but" contradiction statements we define a breakthrough problem. With a little practice you will be able to easily create contradiction statements for any specific technical problem situation or even any technical system or process thus opening the door to finding inventive solutions. For example, create a contradiction for:

A pencil:
- If the lead is solid then it will not break but it won't make a mark on paper.

Vacuum cleaner
- If there is no filter then suction strength will always be strong but particles of dirt will not be captured.

Try creating technical contradiction statements for: a book, a page, a nail.

The above pencil and vacuum cleaner examples show how to create a technical contradiction for a Type 3 general improvement problem for a technical system. If working to solve a specific problem, try to think of what makes the situation better or worse relating to the problem. For example if a filter becomes blocked too quickly, the specific problem is **how to** stop a filter becoming blocked quickly. Try:

- TC-1: If the filter holes are large then the filter does not become blocked but it does not remove particles.
- TC-2: If the filter holes are small, then it filters particles well but it becomes blocked very quickly.

We try to convert any problem situation into a technical contradiction in the form of an "if-then-but" statement which requires an "if" - something that we change, a "then" - something that improves and a "but"- something that worsens. For example, if my tires keep bursting, create: if I increase the thickness of the rubber then the tire does not burst, but the tire becomes too heavy and so on. In this case we simply created the idea that thickness is something we could use as an "if" that relates to ties bursting and weight as a downside to having thick tires.

169

Creating technical contradictions is a key skill that is needed to create breakthrough solutions using TRIZ. **Any fundamental technical problem can be expressed as a technical contradiction**. It is a case of creating an "if-then-but" statement then forming the two conflicts TC-1 and TC-2 that derived from it.

Without trying to form a contradiction we may never be inspired, limiting our solutions to basic incremental or compromise solutions. Trying to form a problem into a contradiction releases psychological inertia (e.g. we might never consider that a bus could be longer because we have a preconceived idea about the length of a bus). Much of the key to innovation is to first define a problem for which a breakthrough is needed.

1.5 What if the problem does not form a contradiction?

For example:
- TC-1: **If** I make the end more pointed **then** nail inserts more quickly **but**.....?

What if we cannot see an opposing downside? If there is no downside then we can simply implement the change, in this case we can go ahead and make the nails end more pointed as our solution. In order to create a breakthrough, we need to solve a contradiction and therefore we must find a downside, something to oppose the change. We could try, if the nail is more pointed then the nail inserts more quickly but the point breaks more easily or the point bends more easily, or when it is too sharp it is dangerous.

Forming the contradiction makes us think of improvements we may not have otherwise considered. So, **if a genuine contradiction can't be found, then make up an artificial one!**

The artificial opposition allows us to form a technical contradiction which is an inventive problem, which we can expose the TRIZ solutions tools for solving contradictions to and so create breakthrough ideas. Even if you make the downside something you may not really see as a negative, the formation of the contradiction statement will still allow us to process the problem using the tools for solving contradictions and therefore help create ideas. As a TRIZ user, it is your role to create or identify contradictions and create inventive solutions by solving them. This is most easily done by using the "if-then-but" statement as a contradiction creation tool.

1.6 Physical Contradiction - One Physical Conflict

A physical contradiction expresses a problem in the form of an opposing physical requirement, something (system, component, part of a component etc.) must be "A" for the good thing to happen but it must be "anti-A" to eliminate the bad thing. So "something" must be "A" and "anti-A," where "A" and "anti-A" are opposite physical requirement, they form antonym pairs for example, hot and cold, large and small, present and absent.

We typically use the statement forms below to state a physical contradiction:

o   To provide the useful action, X must be A, and to eliminate the negative action X must be anti-A. X must be A and anti-A.

o   To provide the useful action X must be A, but it must be anti-A because being anti-A is required. X must be A and anti-A.

For example:

o   To stop rain, an umbrella must be large and to be carried, an umbrella must be small. An umbrella must be large and small.

o   To be transported an iron skillet must be light but it must be heavy because it must be made of cast iron. It must be light and heavy.

In the first example, the X must be A and anti-A to eliminate and support actions. In the second example, X must be A to support an action but also be anti-A because it is an inherent requirement. When forming a physical contradiction we define opposing physical requirements of the same feature, parameter, quality, etc. Those requirements can be to provide or eliminate an action or simply be a necessary requirement for any reason such as in this case, where the case iron skillet must be heavy because by definition it has to be made of cast iron.

"A" and "anti-A" are an antonym pair. For example, something must be hot and cold, big and small, present and non present (being present is a physical state), blue and non-blue, black and white, porous and non-porous, fast and slow, rough and smooth, etc.

Similar to a technical contradiction, a physical contradiction can be thought of as also made up of two physical conflicts, but the conflicts are identical and therefore describe a **single physical conflict**. For example,

State 1: Large Umbrella

o   TC-1: If I increase the size of the umbrella, then it stops rain, but it is difficult to carry.

State 2: Small Umbrella

o   TC-2: If I decrease the size of the umbrella then it is easy to carry but it does not stop rain.

The physical conflict derived from TC-1 is:

o   Physical Conflict-1: To stop rain, an umbrella must be large, but to be carried it must be small. The umbrella must be large and small

The physical conflict derived from TC-2 is:

o   Physical Conflict-2: To be carried, an umbrella must be small, but to stop rain it must be large. An umbrella must be small and large.

In physical conflict-1 the umbrella must be large and small; in physical conflict-2 it must be small and large. The opposing requirements are the same. So we always form only one physical conflict. Since both physical conflicts are the same we will not use the term physical conflict, we will use only the term physical contradiction.

1.7 Why we form Physical Contradictions

Stating the problem as a physical contradiction forces us to consider how to adjust "something" or part of "something" in order to meet the opposite requirements and therefore solve the problem in novel way. Ideally this is simply part of the system or an available resource rather than a new resource that has to be made available at added cost. **Any technical contradiction can be formed into one or more physical contradictions.**

Below when we discuss how to solve contradictions, we will learn that physical contradictions can be resolved in only three ways, by satisfaction, separation or bypass - which focuses our ideas to create breakthrough solutions.

2.   Solving Contradictions Overview

We will break down how to form and solve technical and physical contradictions into **Basic and Advanced levels** (see Contradiction Problem Solving Flowchart – Simplified ARIZ later in this chapter).

We will first learn the basic tools for solving technical and physical contradiction then the advanced tools. The advanced tools are the main problem solving tools of ARIZ and are discussed in detail in Chapter 9 Part 2. Most of those who have learned TRIZ know and apply only the basic tools discussed here.

2.1 Basic Contradiction Problem Solving

The basic method of forming and solving a technical contradiction (the method first created by Altshuller to solve technical contradictions and thus solve inventive problems) is to express the contradiction in terms of **opposing parameters** - using the "if-then-but" statement.

- If (state what action or change could be implemented to create the desired effect)
- Then (state the positive or desired parameter)

- But (state the negative or undesired parameter)

Defining "something" that improves or worsens in terms of parameters, allows us to apply the Contradiction Matrix and 40 inventive principles, the first tool used by Altshuller for solving technical contradictions.

The basic method of forming a physical contradiction is to try to identify antonym pairs (hot and cold, large and small, rough and smooth, etc.) that eliminate the negative and provide the positive parameters (requirements) of the technical contradiction. The physical contradiction is solved using satisfaction, separation and bypass methods which we will discuss shortly.

We state a physical contradiction in the form:

- X-resource should be A to eliminate the negative parameter and anti-A to provide the positive parameter. X-resource must be A and anti-A.

2.2 Advanced Contradiction Problem Solving

Advanced contradiction problem solving methods will be discussed in detail in Chapter 9 part 2. But it is useful to have a basic understanding of them as we describe the basic tools. The advanced methods for technical contradiction problem solving require us to express the technical contradiction in terms of **opposing actions** instead of opposing parameters.
- If (state what action or change could be implemented to create the desired effect)
- Then (state the positive or desired **action**)
- But (state the negative or undesired **action**)

When expressed as opposing actions we can solve the **technical contradiction** using the advanced tools (used in ARIZ):
    1. Su-field modeling and inventive standards.
    2. Space and time resources.
    3. Smart Little People.
    4. An analysis of resources and solution using the Ideal Final Result statement IFR-1 for the technical conflicts.

The advanced level of solving a **physical contradiction** is to use the resource analysis (**Resource Analysis** will be discussed in detail later in Part 2 of this chapter, it is simply a list of all available resources (substances, fields and their parameters) in and around where the problem is that can be used to solve the problem) to systematically create antonym pairs that eliminate and support the negative and positive actions for each resource and thus form many more physical contradictions than the basic method. This allows us to create the Ideal Final Result statement IFR-2 for the physical contradiction at the macro and micro level.

We state the physical contradiction at the advanced level in the form:

- X-resource should be A to eliminate the negative action and anti-A to provide the useful action. X-resource must be A and anti-A.

The physical contradiction is solved using exactly the same tools as the basic methodology; satisfaction, separation and bypass methods. We will discuss the advanced tools in Part 2 of this chapter.

Note that we will use the Contradiction Matrix and the 40 inventive principles as our basic first tool for solving a technical contradiction.

In 1985 Altshuller stopped using the Contradiction Matrix and Inventive Principles and moved instead to using Su-field modeling and inventive standards for solving a technical contradiction (Su-field modeling and inventive standards solve technical contradictions by addressing the negative action of a technical conflict). We will not drop the 40 inventive principles and matrix because they complement the advanced tools – twelve of the solution concepts listed in the 40 inventive principles are not captured by the inventive standards (numbers 2, 6, 7, 8, 9, 10, 12, 14, 17, 21, 27 and 33).

We have combined the basic and advanced tools for solving technical and physical contradictions into a single Contradiction Problem Solving Flowchart – Simplified ARIZ (see later in this chapter).

We recommend the user applies this flowchart for addressing contradictions. Another method is to use ARIZ, the algorithm for solving inventive problems. ARIZ is discussed in Chapter 9 Part 2 and in the Appendix 8.9. It is a sequence of steps for solving an inventive problem (a contradiction).

2.3 ARIZ Overview

ARIZ starts by forming the specific problem to be solved into a technical contradiction and through a sequence of steps, reformulates the problem so as to expose it to the various classical TRIZ tools and specific tools created for the ARIZ algorithm to help solve contradictions. ARIZ does not include the basic tool (the Contradiction Matrix) for solving a technical contradiction.

ARIZ follows a highly structured step-by-step sequence containing nine parts. Each part has individual steps. Many expert users claim only Parts 1-4 are needed for problem solving and that parts 5-9 are "experimental" steps that are "in development." ARIZ is generally considered to be difficult to use. Altshuller believed ARIZ generally requires a minimum of 80 hours of training.

ARIZ-85C
Part 1: Analyze the Problem
Part 2: Analyze the Problem Model (also known as analyze the resources)

174

Part 3: Define the Ideal Final Result and Physical Contradiction
Part 4: Mobilize and use the Substance-Field Resources
Part 5: Apply the Knowledge Base
Part 6: Change or Replace the Problem
Part 7: Review the Solution
Part 8: Apply the Solution
Part 9: Analyze the Problem Solving Process

Our Contradiction Problem Solving Flowchart – Simplified ARIZ leads us through the application of the basic tools and provides the advanced problem solving capability of the key tools used in ARIZ but in a simplified way.

It includes some important new tools e.g. how to analyze a contradiction using a "**Contradiction Diagram**" which is a powerful analytical tool that allows us to break an "if-then-but" technical contradiction statement down into its component parts and reformulate it in terms of opposing actions and opposing parameters. This not only helps us to apply the Contradiction Matrix and inventive principles to the technical contradiction and to form the basic physical contradiction but it also helps us formulate the contradiction into the various formats for applying the ARIZ (advanced) tools for solving technical and physical contradictions.

The basic tools for solving technical and physical contradictions should be used first to solve a contradiction and if needed the advanced tools for solving technical and physical contradictions are applied.

We recommend the user first applies the Contradiction Problem Solving Flowchart – Simplified ARIZ to formulate and solve contradictions (see below section 2.4 and 2.5). If needed, the user may choose to use ARIZ instead of the flowchart (see Appendix 8.9).

2.4 Overview of the Contradiction Problem Solving Flowchart

We have combined the basic and advanced methods for solving technical and physical contradictions into a single Algorithm for solving contradictions (inventive problems). It is a simplified form of ARIZ. It contains four main phases A, B, C and D.

A. Form and Solve the Technical Contradiction using Basic Tools
B. Form and Solve the Physical Contradiction using Basic Tools
C. Form and Solve the Technical Contradiction using Advanced Tools
D. Form and Solve the Physical Contradiction using Advanced Tools

2.5 Contradiction Problem Solving Flowchart – Simplified ARIZ.

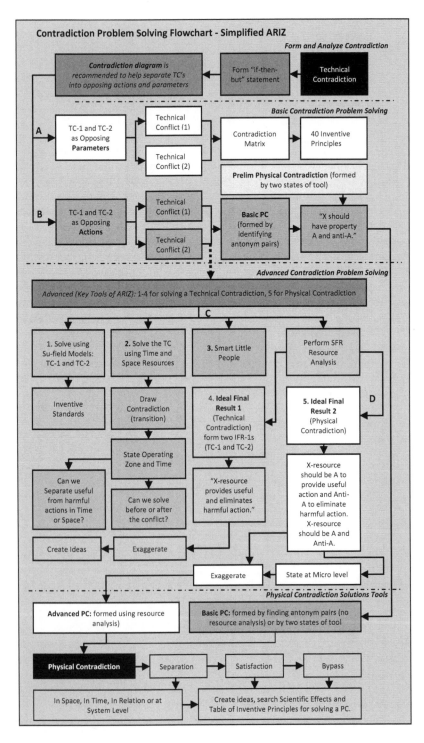

Contradiction Problem Solving Flowchart - Simplified ARIZ

Form and Analyze Contradiction

Contradiction diagram is recommended to help separate TC's into opposing actions and parameters ← Form "if-then-but" statement ← Technical Contradiction

Basic Contradiction Problem Solving

A | TC-1 and TC-2 as Opposing **Parameters** → Technical Conflict (1) / Technical Conflict (2) → Contradiction Matrix → 40 Inventive Principles

Prelim Physical Contradiction (formed by two states of tool)

B | TC-1 and TC-2 as Opposing **Actions** → Technical Conflict (1) / Technical Conflict (2) → **Basic PC** (formed by identifying antonym pairs) → "X should have property A and anti-A."

Advanced Contradiction Problem Solving

Advanced (Key Tools of ARIZ): 1-4 for solving a Technical Contradiction, 5 for Physical Contradiction

C

1. Solve using Su-field Models: TC-1 and TC-2
2. Solve the TC using Time and Space Resources
3. Smart Little People
Perform SFR Resource Analysis

Inventive Standards
Draw Contradiction (transition)
4. **Ideal Final Result 1** (Technical Contradiction) form two IFR-1s (TC-1 and TC-2)
5. **Ideal Final Result 2** (Physical Contradiction)

D

State Operating Zone and Time

Can we Separate useful from harmful actions in Time or Space?
Can we solve before or after the conflict?
"X-resource provides useful and eliminates harmful action."
X-resource should be A to provide useful action and Anti-A to eliminate harmful action. X-resource should be A and Anti-A.

Create Ideas ← Exaggerate ←

Exaggerate ← State at Micro level

Physical Contradiction Solutions Tools

Advanced PC: formed using resource analysis)
Basic PC: formed by finding antonym pairs (no resource analysis) or by two states of tool

Physical Contradiction → Separation → Satisfaction → Bypass

In Space, In Time, In Relation or at System Level → Create ideas, search Scientific Effects and Table of Inventive Principles for solving a PC.

176

Note that Simplified ARIZ the Algorithm for Solving Inventive Problems is a step-by-step process that supports the flowchart is provided in Section 7 of Chapter 9 Part 2.

We will refer to the Contradiction Problem Solving Flowchart as we proceed through the rest of Chapter 9 Parts 1 and 2.

The flowchart maps how we proceed through the analysis of a technical contradiction and apply (A) the basic tools for analyzing and solving a technical contradiction and (B) the basic tools for analyzing and solving a physical contradiction through the advanced tools (C) and (D) for analyzing and solving technical and physical contradictions.

3.   Technical Contradictions Basic

3.1 The Contradiction Matrix and the 40 Inventive Principles

A technical contradiction can be expressed as two technical conflicts. The two technical conflict "if-then-but" statements can be stated in two important ways, as *opposing parameters* and as *opposing actions.*

We form the technical contradiction as two technical conflicts (TC-1 and TC-2) in terms of opposing parameters in order to apply the Contradiction Matrix. The matrix is simply a way of accessing the most frequently used principles that have been used to solve a specific pair of opposing parameters.

We can also state the contradiction in terms of opposing actions. By expressing TC-1 and TC-2 as opposing actions we are able to form the basic physical contradiction, apply ARIZ tools 1-4 (C on the flowchart) and convert the technical contradiction into a physical contradiction for tool 5 (D on the flowchart). We will discuss how to do this after learning how to solve a technical conflict using the Contradiction Matrix and 40 inventive principles and by conversion to a physical contradiction in terms of basic opposing antonym pairs.

For now let's focus on the contradiction expressed as opposing parameters

The first form Altshuller devised for expressing a technical contradiction was to state it in terms of parameters:

o   TC-1: If state 1, then Parameter A is good/desired, but Parameter B is bad/undesired
o   TC-2: If state 2, then Parameter B is good/desired but Parameter A is bad/undesired

For example:

177

o   **TC-1: If** the power is high **then** speed is fast **but** vibration is bad.
o   **TC-2: If** the power is low **then** vibration is bad **but** speed is good.

Power and vibration are parameters. A parameter is a quality, characteristic or feature of a system.

There are a vast number of specific parameters: density, color, mass, velocity, productivity, etc. A technical conflict could be stated as, for example, if density increases, velocity falls, or as speed increases safety falls etc. Altshuller classified all parameters down to a short list of 39 general parameters. We will refer to them as the 39 matrix parameters. To use the Contradiction Matrix, specific parameters must be classified as one of the 39 matrix parameters.

The list of the 39 matrix parameters is given below in the table The 39 matrix parameters and 40 inventive principles and in detail in Appendix 4. Some matrix parameters are very specific like we would use in science or engineering. Others are more qualitative. The matrix is simply a way of accessing the most frequently used principles for a specific pair of opposing parameters.

Note that there are 39 matrix parameters and that there are 40 inventive principles. The numbers 39 and 40 are very close. This proximity is due only to chance. The number of parameters Altshuller could have chosen may have been 50 or 60 etc., but he chose 39. The number of principles may have been more or less depending on how specific a definition is used, Altshuller chose 40. Parameters and Principles are discussed in more detail below.

Let's use a simple diagram to illustrate a technical contradiction

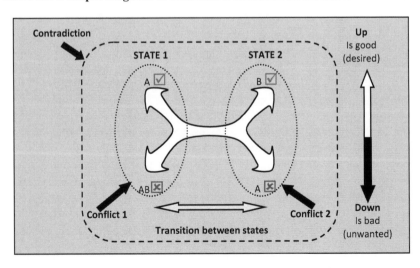

The diagram represents a technical contradiction, which we defined previously as containing two conflicts. In State 1, on the left, we have TC-1 and in State 2 on the right we have TC-2. As we defined previously, a conflict is one half of a contradiction. A conflict is the problem situation in a specific state. The contradiction is represented by both the technical conflict in State 1 **and** the technical conflict in State 2.

The central, four pointed arrow symbol represents the transition as we move back and forward between states. The ☒ and ☑ symbols refer to whether either parameter that we will identify is desired or undesired in that specific state.

Let's take an example: I have a hand-held fan. The handle gets hot when I use too much power.

o   TC-1: If I increase power then speed increases but the temperature increases.
o   TC-2: If I decrease power then the temperature is low but the speed falls.

High speed is good, low speed is bad. High temperature is bad, low temperature is good.

State 1 is high power, State 2 is low power. The change between states is power.

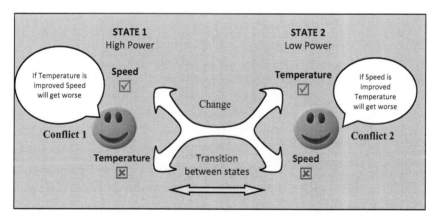

As we move across the diagram and transition from State 1 to State 2, the conflicts reverse (mirror image).

o   In State 1: if we make a change to the right towards State 2 (reduce power), then temperature improves but speed worsens.
o   In State 2: if we make a change to the left State 1 (increase power) then speed improves but temperature worsens.

If the parameter is good in a specific state then it will worsen as we make a change. If it is bad in a specific state then it will improve as we make a change. The "worsening parameter" is the one that is currently good in a specific state, the "improving parameter" is one that is currently bad in a specific state.

o   In State 1: Temperature is the improving parameter, speed is the worsening parameter.
o   In State 2: Speed is the improving parameter, temperature is the worsening parameter.

The parameter that improves or worsens changes depending on the state. So in order to identify an improving or worsening parameter, it is necessary to understand which technical conflict is being referred to.

Altshuller noticed from the study of patents that there were 40 commonly used principles used by inventors. And that these principles could be applied for solving technical contradictions. By classifying contradictions in terms of opposing parameters Altshuller and his colleagues devised a method of helping users find the most frequently used principles to solve specific technical contradiction problems. The principles were classified by improving and worsening parameters. Thus the Contradiction Matrix was created.

Instead of trying to think of new ideas of how to create an inventive solution, the conceptual solutions could be quickly found by using the matrix which guides the user to the inventive principles that others have most frequently used to solve a similar problem in the past. Finding breakthrough solutions could be done quickly and systematically based on data instead of using trial and error or waiting on inspiration.

To use the matrix to solve a technical conflict, identify the matrix parameter that improves and the parameter that worsens. Locate the cell in the matrix for that improving and worsening pair of matrix parameters. Within the cell are listed up to four of the most frequently used inventive principles that were historically used to solve a technical contradiction with those opposing improving and worsening parameters. Prompted by the inventive principles, the user creates ideas to solve technical conflicts using the same concepts that others have used to solve similar conflicts previously.

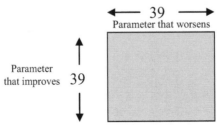

The 39 matrix parameters and 40 inventive principles are shown below. Detailed definitions and examples are included in Appendices 3 and 4.

The 39 Matrix Parameters and 40 Inventive Principles
The table shows a list of the 39 typical parameters that can be in technical conflict (the improving one causes the other to worsen) and the 40 ideas (principles) used to solve a technical conflict. It should be seen as two separate tables, parameter 1 has no relationship to principle 1, etc.

	39 Parameters		40 Principles
1	Weight of moving object	1	Segmentation
2	Weight of stationary object	2	Taking out / Extraction
3	Length of moving object	3	Local Quality
4	Length of stationary object	4	Asymmetry
5	Area of moving object	5	Merging / Combination
6	Area of stationary object	6	Universality
7	Volume of moving object	7	Nested Doll
8	Volume of stationary object	8	Anti-weight / Counter-weight
9	Speed	9	Prior counter-action
10	Force (Intensity)	10	Preliminary action / Prior action
11	Stress or pressure	11	Beforehand cushioning
12	Shape	12	Equi-potentiality / Remove tension
13	Stability of composition	13	The other way round
14	Strength	14	Spheroidality-Curvature
15	Duration of action of moving object	15	Dynamics
16	Duration of action by stationary object	16	Partial or excessive actions
17	Temperature	17	Another Dimension
18	Illumination intensity	18	Mechanical Vibration
19	Use of energy by moving object	19	Periodic action
20	Use of energy by stationary object	20	Continuity of useful action
21	Power	21	Skipping / Hurrying
22	Loss of energy	22	Blessing in Disguise
23	Loss of substance	23	Feedback
24	Loss of Information	24	Intermediary
25	Loss of time	25	Self- Service
26	Quantity of substance	26	Copying
27	Reliability	27	Cheap / Short Living
28	Measurement accuracy	28	Mechanics substitution / Another sense
29	Manufacturing precision	29	Pneumatics and hydraulics / Fluidity
30	Object-affected harmful factors	30	Flexible shells and thin films
31	Object-generated harmful factors	31	Porous Materials / Holes
32	Ease of manufacture	32	Color changes
33	Ease of operation	33	Uniformity
34	Ease of repair	34	Discarding and recovering
35	Adaptability or versatility	35	Parameter changes
36	Device complexity	36	Phase transitions
37	Complexity of control	37	Thermal expansion / Relative change
38	Extent of automation	38	Strong oxidants / Enriched atmosphere
39	Productivity	39	Inert atmosphere / Calmed atmosphere
	----------	40	Composite Materials/ Structures

The Contradiction Matrix

The Contradiction Matrix is a 39x39 matrix used to direct us to the most frequently used of the 40 principles for solving a specific type of conflict where two different parameters are in opposition. The parameter that is to be improved is listed along the side (the horizontal row); the parameter that worsens as a result is listed along the top (in columns)

Below is a detailed section of the matrix. The complete matrix is in Appendix 2.

		Parameter that worsens					
		7. Volume of moving object	8. Volume of Stationary Object	9. Speed	10. Force (intensity)	11. Stress or Pressure	12. Shape
Parameter to be Improved	7. Volume of moving object	PC	-	29, 4, 38, 34	15, 35, 36, 37	6, 35, 36, 37	1, 15, 29, 4
	8. Volume of Stationary Object	-	PC	-	2, 18, 37	24, 35	7, 2, 35
	9. Speed	7, 29, 34	-	PC	13, 28, 15, 19	6, 18, 38, 40	35, 15, 18, 34
	10. Force (intensity)	15, 9, 12, 37	2, 36, 18, 37	13, 28, 15, 12	PC	18,21, 11	10, 35, 40, 34
	11. Stress or Pressure	6, 35, 10	35, 24	6, 35, 36	36, 35, 21	PC	35, 4, 15, 10
	12. Shape	14, 4, 15, 22	7, 2, 35	35, 15, 34, 18	35,10, 37, 40	34,15, 10, 14	PC

The most frequently used principles that have been used to solve the conflict between each pair of improving and worsening parameters are listed in each cell. The inventive principles were identified by studying the patent database. Up to four of the 40 inventive principles are listed in each cell, the most frequently used being at the top left position, less frequently used at bottom right. Where there are less than four principles listed it simply means there were less than four principles that were considered to be statistically significant in terms of frequency of use to solve that technical conflict. If no principles are in the cell it does not mean the principles were not used or the

182

conflict not solved, it means no principles in particular were used; many different ones were used in contrast to a few being used repeatedly.

The center diagonal has no principles indicated. This is where the worsening and parameter to be improved are the same, in those cases a **physical contradiction** is formed (for example, speed has to be fast and speed has to be slow). We don't use the matrix for helping to define or solve physical contradictions. We have a separate methodology for addressing physical contradictions.

You will notice the contents of cells are not symmetrical with respect to the diagonal. This is because the problems in State 1 and State 2 are different. For example, there is a nail being hammered into a block of wood.

o   TC-1: **If** force is high, **then** insertion speed is high (good), **but** stress (bending) is also high (bad).
o   TC-2: **If** force is low, **then** stress is low (good), **but** speed is also low (bad).

At high force, stress is bad, it needs is to be improved and it therefore the improving parameter and speed is good so it is the worsening parameter. At low force speed is bad and needs to be improved, speed is the improving parameter, but stress is good, so it is the worsening parameter. How to reduce stress but maintain high speed when force is high is a different problem from how to increase speed and maintain low stress at low force.

If we try to improve stress but speed worsens the recommended principles to look at first are 6, 35 and 36. If there is an inventive problem where we try to improve speed but stress worsens then 6, 18, 38 and 40 are recommended. The principles are not the same.

What about the Parameter of the Change?

It would be very simple to use the Contradiction Matrix if all we had to consider was - what is the parameter that improves and what is the parameter that worsens when we make a change. We would simply choose the best fit matrix parameters for our specific improving and worsening parameter pairs, and use the matrix to find the appropriate cell and use the principles to create ideas. But it is slightly more complicated than that.

There are often several "best fit" general parameters and we must also consider the parameter that drives the change between states as the "improving" or "worsening" parameter.

When using the matrix, the change between states can also be expressed as a matrix parameter, this is the parameter that drives the change between State 1 and State 2. It can also be considered as the improving or worsening parameter. We'll name this the "parameter of the change."

For simplicity in this example we will use specific parameters that are also matrix parameters, so we can skip the step of converting specific to matrix parameters. The problem is: there is a hand-held fan that gets too hot to hold (handle overheats) when the power is increased. We create the contradiction:

o   TC-1: **If** I increase power, **then** speed increases (good), **but** the temperature increases (bad).
o   TC-2: **If** I decrease power, **then** the temperature drops (good), **but** the speed falls (bad).

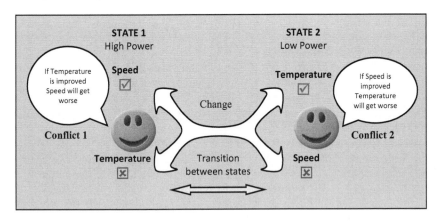

In State 1 speed is good and is the worsening parameter as we transition to State 2, temperature is bad and is the improving parameter as we transition to State 2. In State 2 speed is bad but improves as we transition to State 1 and temperature is good but worsens as we transition to State 1. The parameter that changes between states is power; power is the parameter of the change.

We now have three parameter types:

Specific Parameter: any quality, feature or characteristic that improves or worsens as part of a technical contradiction.

Matrix Parameter: specific parameters are classified as one of 39 matrix parameters.

Parameter of Change: the specific parameter that drives the change between states and then is classified as one of the 39 matrix parameters.

Before we proceed, let's now change the diagram to another format (below) that is easier to draw. Like the illustration above, it shows the improving and worsening parameters, the states and the change between states. There is also a space (inside the boxes) that we will use later to write the improving and worsening actions. We will name it the **"Contradiction Diagram."**

3.2 Basic Contradiction Diagram

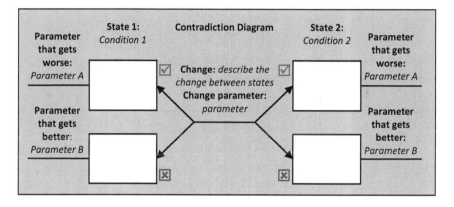

Note that the parameter that gets worse has a tick ☑ because it is good in that state and the parameter that is to be improved has a cross ☒ because it is bad in that state.

State 1 high power:
o TC-1 Temperature is the parameter that improves and speed is the worsening parameter
State 2 low power:
o TC-2 Speed is the parameter that improves and temperature is the worsening parameter.

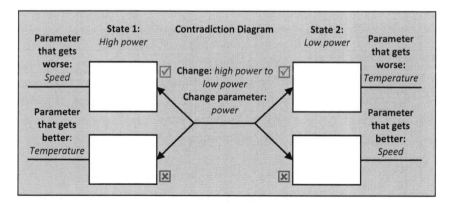

But we also can say in State 1, increased speed *is a consequence of* increased power (even though we may not care if power is high or low, good or bad).

When defining improving and worsening parameters, *the parameter that drives the change between states may also be considered as an improving or worsening parameter.*

In State 1 (high power) we may also have:
- Temperature is the parameter that improves and power is the worsening parameter (because power drives speed)

Because the temperature will fall (get better) as we move to State 2 (low power) but speed will also fall (worsen), and speed is a function of power.

In State 2 (low power)
- Power is the parameter that improves and temperature is the worsening parameter.

The wording used to define a parameter that is improving and one that is worsening can be very confusing. Our definition of improving and worsening parameters is:

o The parameter that improves is also known as the improving parameter, it is bad in its current state and improves (becomes good) as we transition to the opposite state.
o The parameter that gets worse is also known as the worsening parameter, it is good in its current state and worsens (becomes bad) as we transition to the opposite state.

This is why we use the Contradiction Diagram to clarify the parameter that is improving, the parameter that is worsening and the parameter of the change.

In our fan example, we have now defined two sets of improving and worsening parameters in State 1 and State 2. We use the Contradiction Matrix to identify the inventive principles to solve our conflicts.

State 1 High Power, speed is good so it will worsen, temperature is bad (high), it will improve.

But also in State 1, because power drives speed as the parameter of change, we can also say

State 1 High Power, power is good so it will worsen, temperature is bad (high) it will improve.
This gives us for TC-1 (high power state) two cells of the matrix to find inventive principles.

Improving	Worsening	Principles
Temperature (row 17)	Speed (column 9)	2, 28, 36, 30
Temperature (row 17)	**Power (column 21)**	2, 14, 17, 25

186

Similarly for State 2

State 2 Low Power, temperature is good (low) it will worsen, speed is bad (low) and so it is the improving parameter.
But we can also state, because power drives speed:

State 2 Low Power, temperature is good (low) it will worsen, power is bad and so it is the improving parameter.

Improving	Worsening	Principles
Speed (row 9)	Temperature (column 17)	28, 30, 36, 2
Power (row 21)	Temperature (column 17)	2, 14, 17, 25

Principles
        2 Taking out / Extraction
        28 Mechanics substitution / Another sense
        36 Phase transitions
        30 Flexible shells and thin films / Thin & Flexible
        14 Spheroidality
        17 Transition to a new dimension
        25 Self-service

The next step is to apply the principles and create possible solutions (we will complete some more examples below). Also note that it is not necessary to consider each principle individually, it is important to also consider combining principles, for example using principles 2 and 28 together to get a solution.

Prompted by the principles, create solutions:

In State 1(when using high power):

- Principle 2: Taking out or extraction. Remove the hot part of the fan from the handle.

- Principle 25: Self-service. Place the motor (the part that heats) in front of the fan instead of behind it so the fan cools itself.
Example of Principle 25:

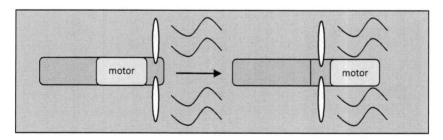

State 2 (when using low power):

- Principle 28: Mechanics substitution/Another sense. Use electrostatic charges of particles in the air driven by an electric field to perform the function of moving air.

- Principle 14: Spheroidality. Continue to use a slow rotating fan but curve and widen the blades to move more air.

Notice that the same principles are suggested (though in a different priority) for each conflict. As we indicated above, this is often not always the case, the principles in each technical conflict are often different (the matrix is not symmetrical). Note that for this contradiction, seven principles out of the total 40 have been suggested.

Many of the parameters are open to a high degree of interpretation. Speed has also been interpreted as productivity (rate of production), or loss of time and so on. In fact there may be a few possible "matrix" parameters that could be chosen for a specific parameter. Given the broad degree of interpretation, the number of principles suggested by the matrix can very quickly increase from four or less, to many more.

Because the principles are concepts classified into a few principles from creative solutions used to solve technical contradictions in the entire researched patent database, they have to be sufficiently general to capture a large number of specific ideas. They therefore provide general concepts of solutions and not very specific ideas.

When solving technical contradictions use the Contradiction Matrix as a guide for finding the most frequently used principles, but consider (stepping through) all the principles and combinations of principles. Let's create an algorithm then review some examples.

Using the Contradiction Matrix and inventive standards is a good starting point for creating solutions to technical contradictions. In our flowchart for solving contradictions we first state the technical contradiction (form an "if-then-but" statement), then use the Contradiction Diagram to analyze the contradiction and then apply the Contradiction Matrix and inventive principles to the two technical conflicts TC-1 and TC-2. Often this "first pass" at applying TRIZ yields many new and inventive ideas.

3.3 Algorithm for Application of Inventive Principles and Contradiction Matrix

Step 1 Form the problem as an "if-then-but" contradiction statement.

Step 2 Form the two "if-then-but" conflict statements, TC-1 and TC-2, and define State 1 and State 2 using the Contradiction Diagram.

Step 3 Define the main useful function of the system. Choose the TC that better delivers the main useful function.

Step 4
A. Identify **specific parameters** that improve or worsen for TC-1 or TC-2
B. Define the improving/worsening **general matrix parameter** pairs for the chosen TC.
C. Identify the parameter of change and form an opposing pair with the parameter of change in place of the improving or worsening parameter it produces.

Step 5 Identify (up to four) principles for each conflicting pair of the chosen TC using the Contradiction Matrix. Propose solutions

For more ideas step through all principles and then try to combine principles or try the other technical conflict.

Examples of Solving Technical Contradictions (using Basic tools)

Example 1: To perform a quality inspection step in the manufacture of decorative plastic film, it is necessary to quality inspect the material for small pinholes which can occur during manufacture. Bright light is used to illuminate the film. The high brightness of the light however damages (decomposes) the material. Let's apply the algorithm.

Step 1 Form the problem as an "if-then-but" contradiction statement.

o If the light is bright then I see the film clearly but the film decomposes.

Step 2 Form the two "if-then-but" conflict statements, TC-1 and TC-2, and define State 1 and State 2 using the Contradiction Diagram.

State 1: Bright light
TC-1: If the light is bright then I see the film clearly but the film decomposes (inspection quality is good, damage is bad).

State 2: Dim light
TC-2: If the light is not bright then the film does not decompose but I cannot see the film clearly (damage is good (none), inspection quality is bad).

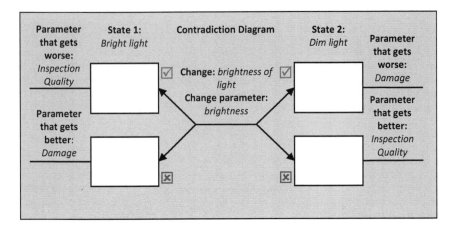

Step 3 Define the main useful function of the system. Choose the TC that better delivers the main useful function.

The main useful function of the system is to inspect. This occurs in TC-1 bright light. For demonstration purposes we will show how to create solutions for both TC-1 and TC-2.

Step 4
A. Identify specific parameters that improve or worsen for TC-1 or TC-2
B. Define the improving/worsening general matrix parameter pairs for the chosen TC.
C. Identify the parameter of change and form an opposing pair with the parameter of change in place of the improving or worsening parameter it produces.

State 1 Bright light
TC-1: If the light is bright then I see the film clearly but the film decomposes (inspection quality is good, damage is bad).

	Improving Parameter	Worsening Parameter	Principles (to be identified in Step 5)
A) Specific Parameters	Damage to film	Inspection quality	
B) Matrix Parameters	Stability (parameter 13)	Accuracy of measurement (parameter 28) Or Loss of information (parameter 24)	
C) Parameter of the Change	Damage to film (parameter 13)	Brightness (parameter 18)	

190

In this case inspection quality is a consequence of brightness, the parameter of the change. It is therefore chosen as the worsening parameter.
State 1 Dim light

TC-2: If the light is dim then the film does not decompose but I can't see the film clearly (damage is good (none), inspection quality is bad).

	Improving Parameter	Worsening Parameter	Principles (to be identified in Step 5)
A) Specific Parameters	Inspection quality	Damage to film	
B) Matrix Parameters	Accuracy of measurement (parameter 28) Or Loss of Information (parameter 24)	Stability (parameter 13)	
C) Parameter of the Change	Brightness (parameter 18)	Damage to film (stability) (parameter13)	

In this case brightness is the improving parameter; inspection quality is a consequence of brightness.

Step 5 Identify the (up to four) principles for each conflicting pair for the chosen TC using the matrix

TC-1:

	Improving Parameter	Worsening Parameter	Principles
A) Specific Parameters	Damage to film	Inspection quality	-
B) Matrix Parameters	Stability (parameter 13)	Accuracy of measurement (parameter 28) Or Loss of information (parameter 24)	13
C) Parameter of the Change	Damage to film (parameter 13)	Brightness (parameter 18)	32,3, 27 and 15

TC-1: In state 1 (when using bright light): Principles 13, 32, 3, 27 and 15 are identified by the matrix as the most commonly used to solve this type of technical conflict.

191

TC-2:

	Improving Parameter	Worsening Parameter	Principles
A) Specific Parameters	Inspection quality	Damage to film	-
B) Matrix Parameters	Accuracy of measurement (parameter 28) Or Loss of Information (parameter 24)	Stability (parameter 13)	32, 35, 13 -
C) Parameter of the Change	Brightness (parameter 18)	Damage to film - stability (parameter13)	32, 3, 27

TC-2: In state 2(when using dim light): Principles 32, 35, 13, 3 and 27 are identified by the matrix as the most commonly used to solve this type of technical conflict.

Principles:
13 The other way round, 32 Color changes, 3 Local Quality, 27 Cheap Disposables, 15 Dynamization, 35 Parameter changes

Propose Solutions
In State 1 (when using bright light): Principles 13, 32, 3, 27 and 15

- Principle 32: Color Changes. Change the color of the light so it does not decompose the film but allows good visibility. Change the color of the material to contrast with the light so it is easily seen.
- Principle 3: Local Quality and Principle 27, Cheap Disposables. Allow the material to be damaged, use a sacrificial measurement location that is not important that can be damaged.
- Principle 15: Dynamization. Inspect many locations but each for a very short time.

In State 2 (when using dim light): Principles 32, 35, 13, 3 and 27

- Principle 13: The other way round. Pass light through the material from behind, a dimmer light can be used to reveal the holes rather than using bright reflected light.

For more ideas step through all principles and then try to combine principles or try the other technical conflict.

Example 2: When I cut a soft tomato, instead of slicing cleanly, the tomato gets squashed.

Step 1 Form the problem as an "if-then-but" contradiction statement.
If I cut a soft tomato quickly with a knife, **then** it gets sliced **but** it also gets squashed.

Step 2 Form the two "if-then-but" conflict statements, TC-1 and TC-2, and define State 1 and State 2 using the Contradiction Diagram.

State 1 High force
  TC-1: If I use high force then I quickly cut through the tomato with a knife but it gets squashed.

State 2: Low force
  TC-2: If I gently force (low force) the knife through the tomato then it doesn't get squashed, but it gets cut slowly.

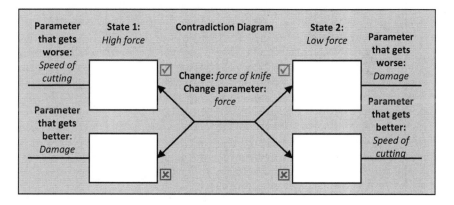

Step 3 Define the main useful function of the system. Choose the TC that better delivers the main useful function.

The main useful function of the system is to cut the tomato. This occurs in TC-1 high force. For demonstration purposes we will consider both TC-1 and TC-2.

Step 4
A. Identify specific parameters that improve or worsen for TC-1 or TC-2
B. Define the improving/worsening **general matrix parameter** pairs for the chosen TC
C. Identify the parameter of the change and form an opposing pair with the parameter of change in place of the improving or worsening parameter it produces.

For TC-1 State 1 High Force
TC-1: If I use high force then I quickly cut through the tomato with a knife but it gets squashed.

	Improving Parameter	Worsening Parameter	Principles (identified in Step 5)
Specific	Damage to Tomato	Speed of Cutting	
Matrix	Stability (13)	Speed (9) Or Loss of Time (25)	
Change	Stability (13)	Force (10)	

In this case Force is the worsening parameter; speed of cutting is a consequence of force.

For TC-2 State 2 Low Force
TC-2: If I gently force the knife through the tomato then it doesn't get squashed, but it gets cut slowly.

	Improving Parameter	Worsening Parameter	Principles (identified in Step 5)
A) Specific	Speed of Cutting	Damage to Tomato	-
B) Matrix	Speed (9) Or Loss of Time(25)	Stability (13)	
C) Change	Force (10)	Stability (13)	

In this case speed of cutting is a consequence of force, the parameter of the change. It is therefore chosen as the improving parameter in part C.

Step 5 Identify the (up to four) principles for each conflicting pair for the chosen TC using the matrix.

TC-1:

	Improving Parameter	Worsening Parameter	Principles
Specific	Damage to Tomato	Speed of Cutting	-
Matrix	Stability (13)	Speed (9) Or Loss of Time (25)	33,15,28,18 35,27
Change	Stability (13)	Force (10)	10, 35,21,16

TC-1: In state 1(when using high force): Principles 33,15,28,18, 35, 27, 10, 21 and 16 are identified by the matrix as the most commonly used to solve this type of technical conflict.

194

TC-2:

	Improving Parameter	Worsening Parameter	Principles
A) Specific	Speed of Cutting	Damage to Tomato	-
B) Matrix	Speed (9)	Stability (13)	28, 33, 1,18
	Or Loss of Time(25)		35,3,22,5
C) Change	Force (10)	Stability (13)	35,10,21

TC-2 In State 2 (when using low force): Principles 28, 33, 1, 18, 35, 3, 22, 5, 35, 10 and 21 are identified by the matrix as the most commonly used to solve this type of technical conflict.

Principles:
28 Mechanical Substitution, 33 Uniformity, 1 Segmentation, 18 Mechanical Vibration, 35 Parameter Changes, 3 Local Quality, 22 Blessing in Disguise, 5 Merging, 10 Preliminary Action, 21 Rushing Through, 15 Dynamization, 27 Cheap Disposables, 16 Partial or Excessive Action, 28 Mechanical Substitution.

Propose Solutions

In State 1(High Force): Principles 33, 15, 28, 18, 35, 27, 10, 21 and 16 are suggested.
- Principle 10 Preliminary Action and Principle 21 Rushing Through. The blade pierces the skin then widens the cut as it proceeds downwards at high speed (rushing through). See illustration below.

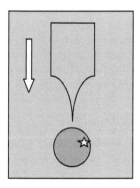

In State 2 (Low Force): Principles 28, 33, 1, 18, 35, 3,22,5, 35, 10 and 21 are suggested.
- Principle 18: Mechanical Vibration. Use low force with vibrating blade to cut fast. Principle 22: Blessing in Disguise. Make a tomato sauce with the squashed tomatoes.

For more ideas step through all principles and then try to combine principles or try the other technical conflict.

Try this exercise; create at least one idea for each principle identified above in Example 2 above for the squashed tomato at high and low force. It will become clear how powerful the inventive principles are at prompting creative thinking.

After the study of millions of patents, Altshuller settled on 39 generic parameters and 40 generic principles. An intensive re-analysis of patents was published in 2003 in a book titled *Matrix 2003 - Updating the TRIZ Contradiction Matrix, Darrell L. Mann, Creax, 2003*. The general conclusion is that there is no need to increase the number of principles. Essentially the same 40 generic principles could be used. It was suggested some additional parameters that improve or worsen should be added including: amount of information, function intensity, noise, harmful emissions, compatibility/connectability, security, safety/vulnerability, aesthetics and control complexity. It was also suggested that a number of the cell contents and sequence of principles changed over time, which is expected as technology develops.

We shall use the standard Contradiction Matrix (see Appendix 2). It has been used effectively for many decades without change. The original matrix is "good enough" given the breadth of interpretation in defining the most suitable parameters and that the matrix is only a guide to which principle to consider first. Stepping through all principles should be performed to improve familiarity with the principles and to cover more than just the most statistically significant prompt.

4. Physical Contradictions Basic

We previously stated that a physical contradiction is formed when a system, component, part of a component etc. (something) has opposing physical requirements.

The system, component, part of a component must have two opposite states, features, parameters etc. X must be "A" and "anti-A" where "A" and "anti-A" are opposite states features, parameters, etc.

o   To provide the useful action X must be "A" and to eliminate the negative action X must be "anti-A." X must be "A" and "anti-A."
o   To provide the useful action X must be "A" but it cannot because it is "anti-A." X must be "A" and "anti-A."

4.1 Some general examples of Physical Contradictions

* An umbrella should be large to stop rain but it must also be small to be carried. Large and small are different values of the same characteristic – size.

196

- Aircraft wheels should be present at take off and not present during flight. Present and not present are two values of the characteristic "to be present."

- A sieve should be porous to allow water to drain but solid to hold the contents. Porous and non-porous are two values of the same characteristic – porosity.

- A cup of coffee should be hot to taste good but it must be cool to be held. Hot and cool are two different values of the characteristic - temperature.

- A cable must be flexible to be coiled but rigid because it has to be made of iron.

It is useful to state why each opposite requirement is needed, as shown in the last example. The requirement might be for any reason; to support an action, to eliminate an action, be required by nature, because we have a lot of iron available and need to use it up, etc.

At the basic level for solving physical contradictions, we do not create a list of resources as we do for advanced level tools, but simply try to identify antonym pairs. This is done by reviewing the technical contradiction. The technical contradiction could be stated as opposing parameter, opposing actions or a mixture of both.

For example:

o TC-1: If an umbrella is big then it stops rain but it is not easy to carry.

o TC-2: If an umbrella small then it is easy to carry but doesn't stop rain.

For TC-1, the umbrella is big, the harmful action is "umbrella is difficult to carry" (ease of use (No. 33) is the harmful parameter), the useful action as we walk along in the rain is "umbrella stops rain" (area of moving object (No. 5) is the useful parameter).

We choose the conflict that best supports the main useful function. In this case it is TC-1. The main useful function of an umbrella is to stop rain which is best delivered when the umbrella is big. As explained previously we need only use one technical conflict, because we produce only one physical contradiction.

Now we identify antonym pairs that support the action or parameters in opposition. We use a simple table.

4.2 Forming Physical Contradictions (Antonym Pairs)

For the chosen technical conflict: in this case the umbrella example above.

A useful action or parameter: (to stop rain) Is a function of:	Anti-A harmful action or parameter : (to be difficult to carry) is a function of :	Parameter, Characteristic or Feature
Big	Small	Size
Wide	Narrow	Width
Present	Absent	Presence
Solid	Porous	Porosity

First we identify items in column 1. We brainstorm what specific parameter, characteristic or feature that the useful action is a function of. In this case the useful action is to stop rain; this is a function of being big, being wide, being present.

Next, in column 2 (anti-A) we identify the antonyms of column 1. Now in column 3 write the specific parameter, characteristic or feature the antonym pair describes. For example big and small are measurements of size, size is input to column 3, wide and narrow are features of width, so input width to column 3. Continue this for all antonym pairs.

Now we input the information from the column into the physical contradiction statement below. If it makes sense then it forms a physical contradiction statement.

The <name the system or object> should be <A> to <state useful action or parameter> and <anti-A> to eliminate <state the harmful action or parameter>. The <name the system or object> should be <A> and <anti-A>.

The< umbrella> should be <**big**> to <stop rain> and <**small**> to eliminate <being difficult to carry>. The <umbrella> should be <**big**> and <**small**>.

The< umbrella> should be <**wide**> to <stop rain> and <**narrow**> to eliminate <being difficult to carry>. The <umbrella> should be <**wide**> and <**narrow**>.

The< umbrella> should be <**present**> to <stop rain> and <**absent**> to eliminate <being difficult to carry>. The <umbrella> should be <**present**> and <**absent**>.

The< umbrella> should be <**solid**> to <stop rain> and <**porous**> to eliminate <being difficult to carry>. The <umbrella> should be <**solid**> and <**porous**>.

The last example does not make sense, being porous is not a feature related to providing the function "to be easy to carry." Only three of these are valid physical contradictions:

198

- The umbrella must be big and small.
- The umbrella must be narrow and wide.
- The umbrella must be absent and present.

This is a simple method of how we form physical contradictions at a basic level. We identify antonym pairs that support the opposing actions/parameters in conflict.

Before we discuss how to solve physical contradictions, let's look at another basic tool for forming physical contradictions.

4.3 Preliminary Physical Contradiction Formation

When we identify the parameter of the change or what changes between the two states of a contradiction, we form a physical contradiction or the "tool." For example:

o TC-1: **If** the bus is long **then** it carries many passengers **but** it can't turn corners.

o TC-2: **If** the bus is short, **then** it can turn corners **but** it doesn't carry many people

The bus (the tool, the tool is the thing that drives the change between states, we will discuss how to identify the tool below in section 5 when discussing the Contradiction Diagram) should be long to carry many passengers and the bus should be short to turn corners. The bus is required to be long and short, which is a physical contradiction. Long and short is the change between State 1 and State 2. We can create a physical contradiction from any technical contradiction in this way. The bus must be long and short, the light must be bright and dim, the force of the hammer must be high and low, etc.

We will discuss how to form physical contradictions from an inventory of available resources later when we discuss the advanced tools. We will call this the Preliminary Physical Contradiction (PPC) because it can be stated prior to resource analysis.

4.4 Why do we form a Physical Contradiction?

The ability to state a technical contradiction as a physical contradiction opens up a number of new ways for solving that contradiction. It is often stated that physical contradictions lead to more innovative solutions than technical contradictions. TRIZ users will try to create and solve the physical rather than technical contradiction.

Our recommendation is to form the technical contradiction ("if-then-but" statement) first then apply the Contradiction Diagram to separate it into parameters and actions. Then the problem can be exposed to other solutions tools and simply converted to a physical contradiction.

So, how do we solve a physical contradiction? We have created what is a seemingly impossible situation. X-resource must be in two opposing states. How do you make something hot and cold, present and not present, large and small? There are three ways a physical contradiction problem can be solved. By **satisfaction**, **bypass** of the problem or use of the **separation** principles.

4.5 How to Solve a Physical Contradiction

We can try to solve the contradiction in only three ways:

- o Satisfaction
- o Bypass
- o Separation
  - o Separation in space
  - o Separation in time
  - o Separation in relation
  - o Separation at system level

Let's define these and look at some examples:

Satisfaction – in some cases it may be possible to satisfy both conflicting needs. The system or component has to be altered to provide the required opposing characteristics. For example, if acid dissolves particles on a surface but also dissolves the surface, the acid must be acidic and non acidic. To solve by satisfaction we could use a "smart" chemical that only dissolves the particles. Solving by satisfaction is highly desirable, but it often requires special "smart" materials or scientific effects (often phase changes can be used – if something has to be solid and liquid, use ice that melts at the point of contact, etc).

Bypass – solve the physical contradiction by moving to a completely different method that avoids the physical contradiction. For example, to connect a computer to the internet, the computer cables must be connected from the modem to the computer but this means damaging walls by making a path for the cable. The cables must be present to transfer information and not present to eliminate damage to the walls. We could bypass the problem if there were no cables to connect. We can transfer information by radio and therefore bypass the contradiction.

In general satisfaction and bypass require significant changes which are less ideal and are often high cost solutions. Satisfaction and bypass solutions are often identified by searching for scientific effects (see Chapter 7 Part 4).

Separation Principles

- o **Separation in space** – a requirement must be present (or high) in one place but not present (or low) at another.
- o **Separation in time** – a requirement must be present (or high) at one time, but not present (or low) at another.
- o **Separation in relation** – a requirement must be present (or high) under one condition and absent (or low) under another.
- o **Separation at system level** – a requirement must be present (or high) at the system level and absent (or low) at the component level (or vice versa). The contradiction is fixed at a subsystem or supersystem level. Can we resolve the conflict by making the whole system have one property but the parts or subsystem has the opposite property?

We can create ideas for how to solve physical contradictions by using our own imagination to think of ways to separate, satisfy or bypass, but it is useful to search scientific effects or borrow ideas from the 40 inventive principles used to solve a technical contradiction. A Table of Specific Inventive Principles to Solve Physical Contradictions is provided below. See also Appendix 7.

Table of Specific Inventive Principles to Solve Physical Contradictions

Satisfaction
Meet opposing requirements, often by the use of a smart material or chemical or scientific effect
36 phase transitions, 37 thermal expansion, 28 mechanical substitution / another sense, 35 parameter change, 38 strong oxidation and 39 inert atmosphere
Bypass
Use a totally different method to bypass the problem often by a scientific effect
25 self service, 6 multi functionality and 13 the other way round
Separation in Space
A characteristic must be present in one place but not present in another
1 segmentation, 2 taking out, 3 local quality, 17 another dimension, 13 the other way round, 14 curvature, 7 nested doll, 30 flexible shells/thin films, 4 asymmetry, 24 intermediary and 26 copying
Separation in Time
A characteristic is present at one time and not present at another.
15 dynamization, 10 prior action, 19 periodic action, 11 beforehand cushioning, 16 partial or excessive action, 21 skipping, 26 copying, 18 mechanical vibration, 37 thermal expansion, 34 discarding and recovering, 9 prior counter action, and 20 continuity of useful action
Separation in Relation
A characteristic must be present for one action and not present for another.
13 the other way round, 35 parameter changes, 32 color changes, 36 phase transition, 31 porous materials, 38 strong oxidants, 39 inert atmosphere, 28 mechanical substitution and 29 pneumatics and hydraulics
Separation at System Level
A characteristic exists at the system level but not at the component level (or vice versa)
Transition to the subsystem (or micro-level)
1 segmentation, 25 self service, 27 cheap disposables, 40 composite materials, 33 uniformity and 12 equipotentiality
Transition to the supersystem
5 merging, 6 universality, 23 feedback and 22 blessing in disguise

4.6 Examples of the ways of Solving Physical Contradictions

Satisfaction

Eyeglass frames must be thin and light to be worn but they are easily bent. The material must be thin and thick. Using principle 36: phase transition (shape memory uses the principle of phase change) the frames are manufactured from shape memory metal that returns them to their original shape.

Satisfaction	
Can the contradictory requirements be solved by satisfaction?	

Try to determine a method to satisfy the problem - search scientific effects, use Table of Specific Inventive Principles to Solve Physical Contradictions.

Bypass

A car key damages the car. The key must be present to open the car but absent to stop damage and wear around the keyhole. Using principle 28: mechanical interaction substitution - the door is unlocked using remote key (electromagnetic waves).

Bypass	
Can the contradictory requirements be solved by bypass?	

Try to determine a method to bypass the problem - search scientific effects, use Table of Specific Inventive Principles to Solve Physical Contradictions

Separation in Space

A cup of coffee must be hot to drink but it must be cool to not burn the hand. Hot and cool are two different values of the parameter - temperature. The opposing requirement is to be hot and cold.

Separation in Space	
Where is there a requirement to be < A >? cool	On the outside of the cup
Where is there a requirement to be < anti-A >? hot	Inside the cup

If there is a requirement to be A and anti-A at the same location then they cannot be separated in space.

Because the requirement to provide the actions are in a different place, a different location in space, they can be separated in space.

We can use principles (and as we will see later Su-field modeling and scientific effects) to help prompt ideas. There are recommended principles to consider for each type of separation, satisfaction and bypass provided in a table above. In this case the opposing requirements to be hot and cold can be separated in space by principle 7: nested doll. If it is a paper cup from a vending machine there are insulating wraps or you can even use a handle.

Separation in Time

An umbrella must be large to stop rain but it must also be small to be carried conveniently when it is not raining.

Separation in Time	
When is there a requirement to be < A >? large	**When it's raining**
When is there a requirement to be < anti-A >? small	**When being carried when it's not raining**

If there is a requirement to be A and anti-A at the same time then they cannot be separated in time.

Folding umbrellas are expanded when it's raining and folded when the rain stops. The useful function and harmful function are separated in time. They are separated by using Principle 15: Dynamization.

203

Separation in Relation

A pot must be solid to hold potatoes but porous to drain water.

Separation in relation	
For what is there a requirement to be < A >? porous	passing water
For what is there a requirement to be < anti-A >? solid	stopping potatoes

If there is a requirement to be A and anti-A for the same action or object then they cannot be separated in relation.

Porous and solid have two values of the same characteristic – porosity. The sieve separates the water from the potatoes. In relation to the potatoes the sieve is solid. In relation to the water the sieve is porous. Conflicting requirements are separated in relation. They are separated using Principle 31: Porous materials.

Separation at System Level

Opposing requirements can be resolved at one level and eliminated at another. A bicycle chain must be solid to pull the wheel but it must be flexible to move in a circular motion. The chain must be solid and flexible.

Separation at system level (parts v whole)	
For what level of the system is there a requirement to be < A >? **flexible**	**System level**
For what level of the system is there a requirement to be <anti- A >? **solid**	**Micro/subsystem level**

If there is a requirement to be A and anti-A at all different system levels then they cannot be separated at system level.

The chain is solid at the micro/subsystem level yet flexible as a whole Principle 1: Segmentation is demonstrated.

OR

The hull of a boat must be narrow to move quickly through the water but it must be wide to achieve balance. A two-hulled boat is narrow at the system level but wide at the supersystem level (one hull merged with another hull, simplest transition to the supersystem via the trend of mono-bi-poly). This is an example of separation of system and supersystem level. The conflict is solved by Principle 5: Merging.

No matter which source we use to form a physical contradiction: via the Preliminary Physical Contradiction, forming basic physical contradiction statements by creating antonym pairs, or advanced tools to form antonym pairs by resource, we solve it the same way by separation, satisfaction or bypass.

4.7 Template for Solving a Physical Contradiction by Separation, Satisfaction and Bypass

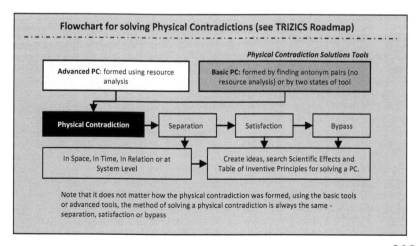

When the physical contradiction has been defined it is useful to use the following template to solve by satisfaction, bypass and separation.

Template for Solving a Physical Contradiction by Separation, Satisfaction and Bypass

Satisfaction	
Can the contradictory requirements be solved by satisfaction?	

Try to determine a method to satisfy the problem - search scientific effects, use Table of Specific Inventive Principles to Solve Physical Contradictions.

Bypass	
Can the contradictory requirements be solved by bypass?	

Try to determine a method to bypass the problem - search scientific effects, use Table of Specific Inventive Principles to Solve Physical Contradictions.

Separation in Space	
Where is there a requirement to be < A >?	
Where is there a requirement to be < anti-A >?	

If there is a requirement to be A and anti-A at the same location, then they cannot be separated in space.

Separation in Time	
When is there a requirement to be < A >?	
When is there a requirement to be < anti-A >?	

If there is a requirement to be A and anti-A at the same time, then they cannot be separated in time.

Separation in relation	
For what is there a requirement to be < A >?	
For what is there a requirement to be < anti-A >?	

If there is a requirement to be A and anti-A for the same action or object, then they cannot be separated in relation.

Separation at system level (parts v whole)	
For what level of the system is there a requirement to be < A >?	
For what level of the system is there a requirement to be <anti- A >?	

If there is a requirement to be A and anti-A at all different system levels, then they cannot be separated at system level.

The steps for creating and solving a physical contradiction are formed into a three-step algorithm in section 6, page 221 below - Physical Contradiction Algorithm using Basic Tools (3 Steps)

5. **The Contradiction Diagram** (How to analyze a Technical Contradiction and express it as opposing actions and as opposing parameters)

Above, we discussed how to state a technical contradiction as two technical conflicts in terms of opposing parameters. The parameters are fitted into a group of 39 general "matrix" parameters that direct us to the most frequently used of the 40 inventive principles. The solutions suggested are general concepts based on how others (from the patent database), have solved the problem. The principles are general concepts of solutions.

If we define our specific problem as opposing actions we can develop more specific solutions relevant to the problem (instead of principle concepts), and we are able to translate the contradiction problem into formats that allow us to apply the tools of ARIZ.

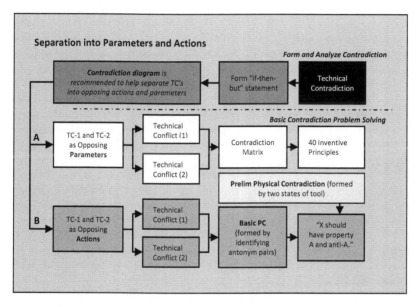

The simplest way to analyze a technical contradiction is to break it down into its component parts using the Contradiction Diagram in parallel with the "if-then but conflict" statements.

We will use the Contradiction Diagram below as a visual aid to deconstruct a technical contradiction, simplify it by breaking it into its component parts then and convert it into opposing actions and opposing parameters. We identify the tool-action-object (or object-action-tool) useful (desired) and harmful (undesired) actions and create a *Graphical Representation* of the conflict in each state. Inventive standards are used to improve the undesired action in either state to provide solutions.

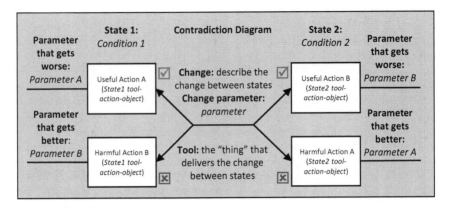

The diagram above is similar to the basic one we used before, as we transition across between states, parameters either improve or worsen. The actions associated with the parameters will also improve or worsen. The information relating to action will be input into the boxed areas. The information relating to parameters will be shown alongside the boxes. We will also identify the change parameter and the tool and states.

Using the diagram is a bit like playing the game **Sudoku**. From the "if-then-but" statement we can deduce everything else, the two states, define the parameter of change, the "tool" which delivers the change. We can then express the two technical conflicts, as opposing parameters or as opposing (tool-action-object) actions.

5.1 Algorithm for Solving Technical Contradictions using the Basic Tools (inventive principles and the contradiction matrix and inventive standards)

Step 1. State the "if-then-but" contradiction statement and define the technical conflict statements for State 1 and State 2.

Step 2. Identify the change between states TC-1 and TC-2.

Step 3. Identify the tool (the thing, not a field that delivers the change between states).

Step 4. Identify the useful and harmful actions for TC-1 and TC-2 in terms of <State 1>Tool–Action–Object, and <State 2>Tool-Action-Object and input them into the diagram.

Step 5. Identify the improving and worsening matrix parameters for both states.

Step 6. Sanity-check the diagram, TC-1 and TC-2 should be the mirror image of each other.

Step 7. Make Graphical Representations of TC-1 and TC-2.

Step 8. For the Graphical Representations, ensure the interaction between the tool and object(s) is direct.

Step 9. Solve the technical conflicts using inventive standards.

208

10. Solve the technical conflicts using the Contradiction Matrix and inventive principles.

For the technical contradiction: "**If** the chemical is strong, **then** it removes more dirt **but** it damages the surface," complete the analysis of the technical contradiction using the Contradiction Diagram.

Step 1. State the "if-then-but" contradiction statement and define the technical conflict statements for State 1 and State 2.

State 1: Weak Chemical

o   TC-1: **If** the chemical is weak, **then** the surface is not damaged, **but** it does not remove dirt well.

State 2: Strong Chemical

o   TC-2: **If** the chemical is strong, **then** it removes more dirt, **but** it damages the surface.

State 1 (weak chemical) If A the cleaning effectiveness is improves, then the damage B will increase.
State 2 (strong chemical) If B the damage improves then A the cleaning effectiveness will get worse.

To create an action statement we need three elements, subject, action and object. We will form our action statements for contradictions in relation to the subject that delivers the change between states. The convention in TRIZ is to name this the "tool" when dealing with a contradiction. The contradiction is stated in terms of tool-action-object statements where the tool is the physical object that delivers the change. This allows us to describe the contradiction in terms of how the tool changes the situation in State 1 to State 2 by describing how the tool interacts differently with the object (or objects) in both states.

In the diagram, we add information about what the change is and the tool that delivers the change.

Step 2. Identify the change between states TC-1 and TC-2: chemical change from weak to strong. The change parameter is therefore *chemical strength*, the state describing State 1 is *weak chemical*. The description of State 2 is *strong chemical*.

Step 3. Identify the tool (the thing, not a field that delivers the change between states): *Chemical*

We input this information into the diagram.

209

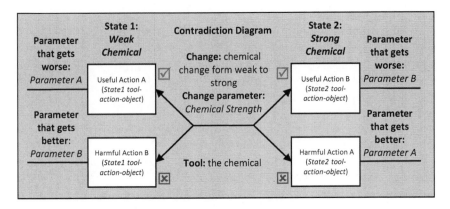

Step 4. Identify the useful and harmful actions for TC-1 and TC-2 in terms of <State 1> tool–action–object, and <State 2> tool-action-object and input them into the diagram.

When naming the tool, the tool **must be preceded with a description of the state** otherwise the action can be confused with what is occurring in the opposite state. The tool should be described as State 1 tool and State 2 tool. For example, *high strength* chemical, *low strength* chemical or *high power* engine, *low power* engine, instead of just chemical and engine.

Referring to TC-1 and TC-2

o TC-1: **If** the chemical is weak, **then** it does not damage the surface, **but** it does not remove dirt well.
o TC-2: **If** chemical is strong, **then** it removes more dirt, **but** it damages the surface.

We derive the useful and harmful actions in each state (we will refer to them as useful and harmful in general, but for a harmful action we may also identify the type of harmful action as insufficient, excessive, absent etc). From our TC-1 and TC-2 statements we identify the object or objects affected by the tool in each technical conflict.

The tool is the (strong - weak) chemical and the objects are the surface and the dirt.

State 1:
Useful Action A: **weak chemical** does not damage **surface** (useful)
Harmful Action B: **weak chemical** does not remove **dirt** well (insufficient)

State 2:
Useful Action B: **strong chemical** effectively removes **dirt** (useful)
Harmful Action A: **strong chemical** damages **surface** (harmful)

We input these actions into the diagram.

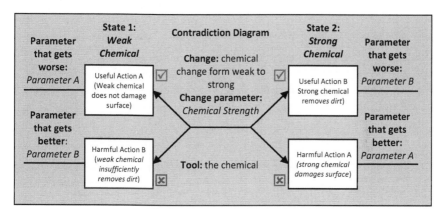

Step 5. Identify the improving and worsening matrix parameters for both states.

The parameters reflect what the actions are doing and vice versa. Steps 4 and 5 are done together. We input the parameters.

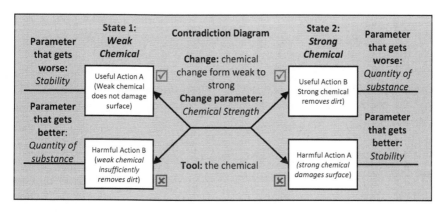

Step 6. Sanity-check the diagram. TC-1 and TC-2 should be the mirror image of each other.

Make sure when chemical is strong the positive and negative actions correctly describe the situation. When the chemical is weak the positive and negative actions are also correct.

The actions are simple interactions that can be easily converted to a simple **Graphical Representation** of the technical conflict in each state. A

Graphical Representation is like a very basic functional model of the two technical conflicts.

Step 7. Make Graphical Representations of TC-1 and TC-2

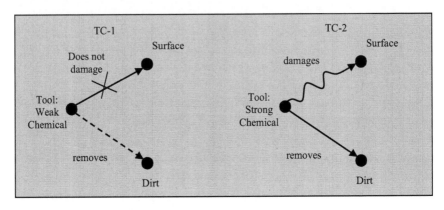

Tool: Chemical, Object 1: Surface, Object 2: Dirt

We have translated our technical contradiction into two Su-field models! We can solve technical contradictions using Su-field models and the 76 inventive standards instead of using the Contradiction Matrix. Note that this is an ARIZ tool, but we will include it in our algorithm for solving technical contradictions using basic tools - Simplified ARIZ. We can apply inventive standards to the harmful (absent, insufficient, excessive) interactions to solve each conflict.

In State 1 improving the insufficient action between weak chemical and dirt solves the problem. Removing the harmful action between strong chemical and the surface also solves the problem. Solutions are provided by the inventive standards. Altshuller preferred this method (inventive standards) over the matrix (inventive principles) because the solutions are more directly relevant to the problem (less conceptual).

In this case the useful action in TC-1 (weak chemical does not damage surface) is a null positive action. A null positive action is an action that is deemed useful because it "doesn't do a bad thing." A normal useful action is useful because it performs a desired action.

Step 8. For the Graphical Representation, ensure the interaction between the tool and object(s) is direct.

See Chapter 8 for Algorithm of Applying Inventive Standard Solutions. It is important to validate the interaction between tool and object(s) are direct. This is because we are forming a Su-field where the tool (S2) must act directly on the object (S1) via a field.

212

Check there are no intermediate objects and functions that link between tool and object (S2 and S1).

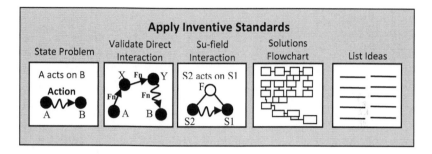

Step 9. Solve the technical conflicts using inventive standards.

Improve the harmful action in both TC-1 and TC-2 (where by harmful we mean absent, harmful, excessive or insufficient). Note, this example is solved in Part 2 of this chapter.

Step 10. Solve the technical conflicts using the Contradiction Matrix and inventive principles.

Solution using the Contradiction Matrix:

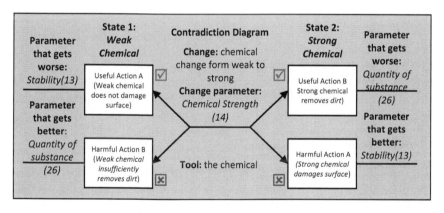

State 1 Weak Chemical (quantity of dirt to be improved, strength to be improved, but stability (decomposes) worsens)

	Improving	Worsening	Principles
B) Matrix Parameters	Quantity of Substance (row 26)	Stability (column 13)	15,2,17,40
C) Parameter of Change	Strength (row 14)	Stability (column 13)	13,17,35

State 2 High Strength (stability to be improved, strength to worsen, quantity of substance to worsen).

	Improving	Worsening	Principles
B) Matrix Parameters	Stability (row 13)	Quantity of Substance (column 26)	15, 32,35
C) Parameter of Change	Stability (row 13)	Strength (column 14)	17, 9, 15

Inventive Principles are: 15 Dynamization, 2 Extraction, 17 Transition to new dimension, 40 Composite Materials, 13 The other way round, 35 Transformation Properties, 32 Color change and 9 Prior counteraction

Solutions: dynamize, add vibration to the weak chemical to improve removal of dirt and which will not damage the surface.

5.2 Examples of Algorithm for Solving Technical Contradictions using the Basic Tools (inventive principles and the contradiction matrix and inventive standards)

Example 1
The problem: If the strength of the chair increases its weight gets worse.

Step 1. State the "if-then-but" contradiction statement and define the technical conflict statements for State 1 and State 2.
o   TC-1 If the chair is strong then its strength increases but its weight gets worse.
o   TC-2 If the chair is weak then its weight is low but its strength gets worse.

Now we must use the Contradiction Diagram to make sense of the problem in terms of parameters and useful and harmful Tool-Action-Object actions

To be able to apply each of the problem solving tools we need to reveal the positive and negative actions, the positive and negative parameters that reflect those actions, define the two states, the tool that delivers the change between states 1 and 2 and the objects affected by the change in the tool between states 1 and 2.

Step 2. Identify the change between states TC-1 and TC-2: strength, low to high strength chair.

Step 3. Identify the tool (the thing, not a field that delivers the change between states): chair.
The chair is the physical object that delivers the change between states.

214

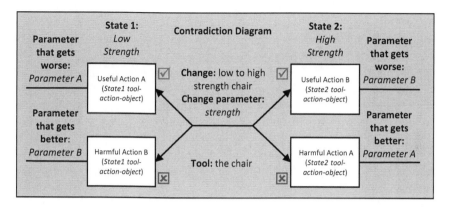

Now we have to use the diagram to help complete the rest of the analysis. This is where we deduce the rest of the information from the basic information of the "if-then-but" statement in order to complete the entire technical Contradiction Diagram.

Step 4. Identify the useful and harmful actions for TC-1 and TC-2 in terms of <State 1> tool–action–object, and <State 2> tool-action-object and input them into the diagram.

See 2nd diagram below

Step 5. Identify the improving and worsening matrix parameters for both states. We can do this before actions are deduced.

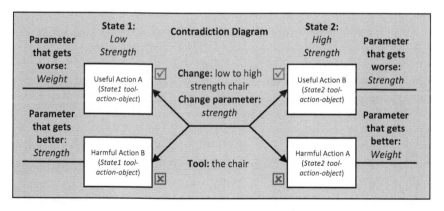

What is good about a chair being strong? It holds people better. What is bad about it being weak? It holds people worse. Similarly what is good about low weight? It is easier for a person to transport or move. What is bad about high weight is it is more difficult to move.

Let's then complete the diagram

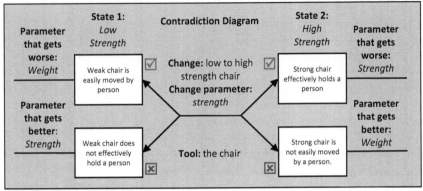

So in terms of actions
- TC-1: **If** the chair is weak **then** it is easily moved by a person **but** does not effectively hold a person.
- TC-2: **If** the chair is strong **then** it effectively holds a person **but** it is not easily moved by a person.

Step 6. Sanity-check the diagram, TC-1 and TC-2 should be the mirror image of each other.

The diagram is correct, the actions relate to the parameters and each side of the contradiction and they form a mirror image.

7. Make Graphical Representations of TC-1 and TC-2.

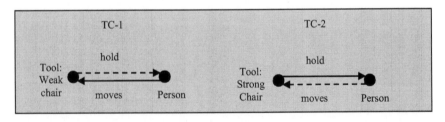

Tool: chair, Object 1: person, Object 2: none

In this case there is only one object in the contradiction. The tool is the chair, the object is person. In the previous example, the chemical was the tool and there were two objects, the surface and the dirt. When a contradiction is formed into a basic interaction (Graphical Representation) there is one tool and either one or two objects, no more.

Step 8. For the Graphical Representations, ensure the interaction between the tool and object(s) is direct.

216

Tool (chair) interacts directly with the person; therefore the tool is correctly defined. We have created two Su-field problems. Note that this time we only have one object that interacts with the chair (a person). When we convert a contradiction into actions, there is one tool and either one or two objects. "Something" (the tool) delivers a change between states and something gets better (an action between the tool and an object) and something else gets worse (an action between the tool and the same or a different object).

Step 9. Solve the technical conflicts using Inventive Standards.

Use inventive standards to create solutions for both technical conflicts.

Step 10. Solve the technical conflicts using the Contradiction Matrix and inventive principles.

Use the Contradiction Matrix and 40 inventive principles to identify solutions for both technical conflicts.

Let's only complete the technical Contradiction Diagram and Graphical Representations for these next two examples:
Example 2: If I increase the force of a hammer blow the nail inserts quicker but the nail bends

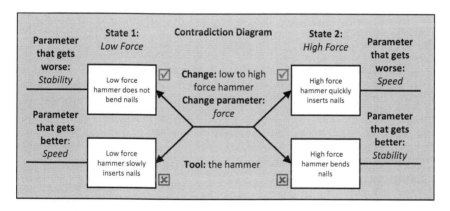

Graphical Representation

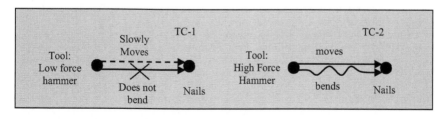

Tool: hammer, Object 1: nails, Object 2: none

217

We can solve both technical conflicts by using inventive standards to improve the harmful action of each conflict or by using the Contradiction Matrix and 40 inventive principles.

Example 3: If the power of the plasma increases, the film is removed vertically but too much resist erosion occurs and the plasma breaks through the resist and removes some of the metal the film that sits beneath the resist mask.

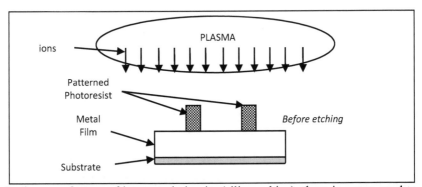

In the manufacture of integrated circuits (silicon chips), there is a step used to pattern metal interconnect between parts of the device. The metal interconnect is similar to wire that conducts electrons but at a microscopic level. To make the "wiring," an entire silicon wafer that contains many devices is completely covered in a thin layer of the conductive film of uniform thickness. The film must be removed in areas where it is not needed and the remaining metal forms a wire pattern that electrically connects parts of the device. To remove unwanted metal, a photosensitive rubber-like material (photoresist) is deposited uniformly across the wafer. It is then exposed to light using a patterned mask that stops light from reaching areas where the metal film is to be removed. The areas that were not exposed to light are removed in a chemical solution leaving a patterned photoresist mask. The photoresist stops the metal underneath it from being removed (etched). To remove exposed metal that is not protected by the photoresist, a process called Reactive Ion Etching is used. This involves creating plasma in a low pressure chamber and accelerating chemically reactive ions toward the patterned metal.

State 1 Low Power
TC-1: **If** the plasma power is low, **then** resist is removed at an effective rate, **but** there is non-vertical etching of the metal film.

State 2 High Power
TC-2: **If** the plasma power is high, **then** there is vertical etching of the metal film, **but** resist is removed excessively.

218

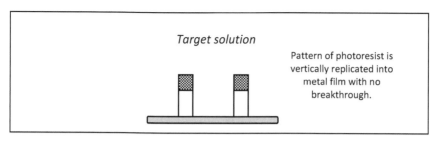

Target solution

Pattern of photoresist is vertically replicated into metal film with no breakthrough.

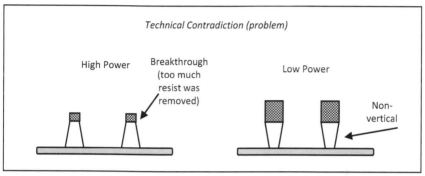

Technical Contradiction (problem)

High Power

Breakthrough (too much resist was removed)

Low Power

Non-vertical

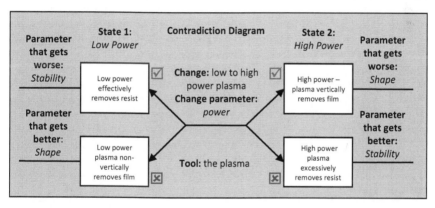

Contradiction Diagram

Parameter that gets worse: *Stability*

State 1: *Low Power*

Low power effectively removes resist ☑

State 2: *High Power*

Parameter that gets worse: *Shape*

Change: low to high power plasma ☑
Change parameter: *power*

High power – plasma vertically removes film

Parameter that gets better: *Shape*

Low power plasma non-vertically removes film ☒

Tool: the plasma

High power plasma excessively removes resist ☒

Parameter that gets better: *Stability*

Graphical Representation

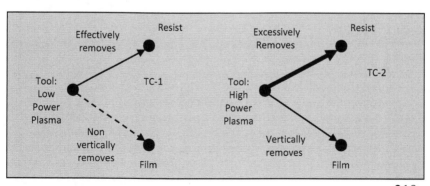

Effectively removes — Resist

Tool: Low Power Plasma

TC-1

Non vertically removes — Film

Excessively Removes — Resist

Tool: High Power Plasma

TC-2

Vertically removes — Film

219

Tool: plasma Object 1: resist Object 2: film

For the Graphical Representation, ensure the interaction between the tool and object(s) is direct.
Does the plasma directly interact with the resist and film? Let's draw our basic Functional Model between the plasma (tool) and resist and film (objects).

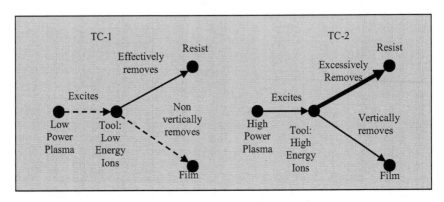

Although the power of the plasma does change the situation from State 1 to State 2, the ions themselves perform the useful and harmful functions - not the plasma. It is important to validate that the problem is correctly formulated by creating a simple functional model of the problem between tool and objects. The tool is the ions, not the plasma.

Building a functional model reveals the plasma excites ions, the ions accelerate towards the resist and film. A simple functional model is useful for checking the problem is correctly defined.

Tool: ions, Object 1: resist, Object 2: film

6. Algorithm for solving Technical and Physical Contradiction using Basic Tools (part A of Simplified ARIZ)

We combine the algorithm for forming and solving a technical contradiction (10 steps) with basic methods form formulating and solving a physical contradiction (3 steps). First form and solve the technical contradiction, then form and solve the physical contradiction.

Technical Contradiction Algorithm using Basic Tools (10 Steps)
Step 1. State the "if-then-but" contradiction statement and define the technical conflict statements for State 1 and State 2.
Step 2. Identify the change between states TC-1 and TC-2.
Step 3. Identify the tool (the thing, not a field that delivers the change between states

220

Step 4. Identify the useful and harmful actions for TC-1 and TC-2 in terms of <State 1>Tool–Action–Object, and <State 2>Tool-Action-Object and input them into the diagram.
Step 5. Identify the improving and worsening matrix parameters for both states.

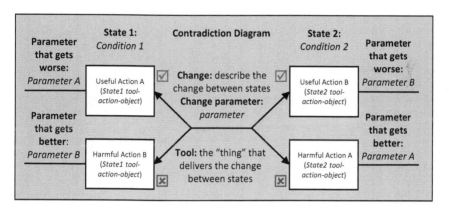

Step 6. Sanity-check the diagram, TC-1 and TC-2 should be the mirror image of each other.
Step 7. Make Graphical Representations of TC-1 and TC-2
Step 8. For the Graphical Representations, ensure the interaction between the tool and object(s) is direct.
Step 9. Solve the technical conflicts using inventive standards.
Step 10. Solve the technical conflicts using the Contradiction Matrix and inventive principles

Note that solving technical contradictions using Su-field modeling and inventive standards is an advanced tool of ARIZ but we will use it in our algorithm for solving contradictions using basic tools.

Physical Contradiction Algorithm using Basic Tools (3 Steps) (part B of simplified ARIZ)

Step 1. For the chosen technical conflict, identify antonym pairs that support the action or parameters in opposition. We use a simple table.

For the chosen Technical Conflict:

A useful action or parameter: Is a function of:	Anti-A harmful action or parameter : is a function of :	Parameter, Characteristic or Feature

Identify items in column 1. brainstorm what specific parameter, characteristic or feature that the useful action is a function of.

Next, in column 2 (anti-A) we identify the antonyms of column 1. Then in column 3 write the specific parameter, characteristic or feature the antonym pair describes.

Now we input the information from the column into the physical contradiction statement below. If it makes sense then it forms a physical contradiction statement.

The <name the system or object> should be <A> to <state useful action or parameter> and <anti-A> to eliminate <state the harmful action or parameter>. The <name the system or object> should be <A> and <anti-A>.

Compile a list of physical contradictions and solve using the methods of satisfaction, separation and bypass in Step 3 (below).

Step 2. Form the Preliminary Physical Contradiction Formation

When we identify the parameter of the change or what changes between the two states of a contradiction, we form a physical contradiction or the "tool."

For example:

o   TC-1: **If** the bus is long **then** it carries many passengers **but** it can't turn corners.
o   TC-2: **If** the bus is short, **then** it can turn corners **but** it doesn't carry many people

The Preliminary Physical Contradiction is: The <bus> should be < long> and <short>.

The <tool> should be <state 1> and <state 2>

And solve using Step 3.

Step 3. Solve the Physical Contradiction

When the physical contradiction has been defined it is useful to use the template for solving a physical contradiction by separation, satisfaction and bypass

Satisfaction	Solution
Can the contradictory requirements be solved by satisfaction?	

Try to determine a method to satisfy the problem - search scientific effects, use Table of Specific Inventive Principles to Solve Physical Contradictions.

Bypass	Solution
Can the contradictory requirements be solved by bypass?	

Try to determine a method to bypass the problem - search scientific effects, use Table of Specific Inventive Principles to Solve Physical Contradictions.

Separation in Space	Where
Where is there a requirement to be < A >?	
Where is there a requirement to be < anti-A >?	

If there is a requirement to be A and anti-A at the same location, then they cannot be separated in space.

Separation in Time	When
When is there a requirement to be < A >?	
When is there a requirement to be < anti-A >?	

If there is a requirement to be A and anti-A at the same time, then they cannot be separated in time.

Separation in relation	For what or whom?
For what is there a requirement to be < A >?	
For what is there a requirement to be < anti-A >?	

If there is a requirement to be A and anti-A for the same action or object, then they cannot be separated in relation.

Separation at system level (parts v whole)	Level
For what level of the system is there a requirement to be < A >?	
For what level of the system is there a requirement to be <anti- A >?	

If there is a requirement to be A and anti-A at all different system levels, then they cannot be separated at system level.

At this point, many of those who study and use TRIZ simply use the basic tools above to solve technical contradictions. Using the Algorithm for solving Technical and Physical Contradiction using Basic Tools is sufficient for them to create and solve contradictions and hence create breakthrough solutions.

However, for more difficult inventive problems, the advanced tools of ARIZ are useful. Next in Chapter 9 Part 2 we learn those advanced tools. Not only will this allow the user to apply the tools of ARIZ as discrete tools, but it will enable the user to better understand the steps and processes used in ARIZ. See Appendix 8.9 for ARIZ-85C.

Chapter 9

Contradictions and ARIZ Tools

Part 2

Advanced Contradiction Problem Solving

Chapter Contents

Advanced Tools for Solving Technical Contradictions

1. **Advanced Tool** 1: Su-field Modeling and Inventive Standards
   1.1 Convert the Technical Contradiction into Su-field Models and apply Inventive Standards
   1.2 Algorithm for Applying System of Inventive Standard Solutions to a Technical Contradiction
2. **Advanced Tool 2**:Solutions in Time and Space (Technical Contradiction)
   2.1 Solving the Technical Contradiction using Time and Space Resources Algorithm
      o Examples
3. **Advanced Tool 3:** Use Smart Little People to solve a Technical Contradiction
   3.1 Algorithm for SLP
4. Substance-Field Resource Analysis (required for Tools 4 and 5)
   4.1 Example of Substance-Field Resource Analysis Table
   4.2 Example of SFR list creation
   4.3 Secondary Resources
5. **Advanced Tool 4:** The Ideal Final Result 1 (IFR-1)
   5.1 State the Technical Contradiction in the form of the Ideal Final Result 1 (IFR-1)
      o Example
   5.2 Algorithm for IFR-1

Advanced Tools for Solving Physical Contradictions

6. **Advanced Tool 5:** Solve the Physical Contradiction by SFR Resource (IFR-2)
   6.1 Convert the Technical Contradiction to a Physical Contradiction (IFR-2)
   6.2 Algorithm for solving Physical Contradictions by Resource (IFR-2)

Simplified ARIZ

7. **Simplified ARIZ the Algorithm for Solving Inventive Problems**

Chapter Summary: *See Chapter 9 Part 1, it contains the summary of Chapter 9 Part 1 and Part 2 combined.*

1. Advanced Tool 1: Su-field Modeling and Inventive Standards

The flowchart for solving contradictions directs us first to find solutions to contradictions using basic tools and then advanced tools. We discussed basic tools in Chapter 9 Part 1. If a satisfactory solution is not identified using basic tools, then use the advanced tools. The advanced tools are five key tools used in ARIZ the algorithm for solving inventive problems (see Appendix 8.9). Instead of applying them via ARIZ, we can apply them as the individual tools described below.

Basic
1. Form the technical contradiction as two technical conflicts in terms of opposing parameters and solve them using the Contradiction Matrix and 40 inventive principles.
2. Form physical contradictions by finding antonyms pairs and solve using separation, satisfaction and bypass.

Advanced
Apply ARIZ tools for creating and solving a technical contradiction:
Tool 1: Apply Su-field modeling and inventive standards to the Graphical Representations of the technical conflicts expressed as opposing actions.
Tool 2: Solutions in Time and Space.
Tool 3: Apply Smart Little People to the technical contradiction Perform Substance-Field Resource Analysis.
Tool 4: Create and solve the IFR-1 statement.
Apply ARIZ tools for creating and solving a physical contradiction:
Tool 5: Form physical contradictions by finding antonym pairs for the resources and solve using separation, satisfaction and bypass.

1.1 Convert the Technical Contradiction into Su-field Models and apply Inventive Standards

We discussed this in Part 1. Although an advanced tool of ARIZ, we included it as a basic tool in the algorithm for solving a technical contradiction. From the Graphical Representations we have defined Su-fields and can therefore apply inventive standards. Let's return to the example 1 that we used in Part 1; the two states were a strong and weak chemical:

226

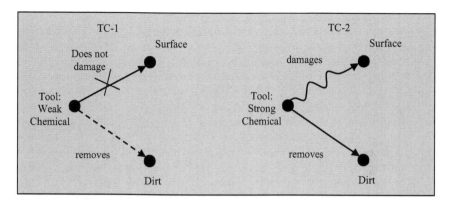

1.2 Algorithm for Applying System of Inventive Standard Solutions to a Technical Contradiction

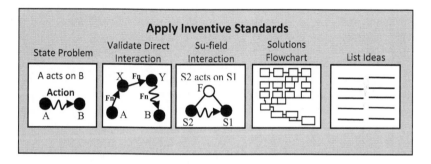

Step 1 State problem as Tool-Action-Object statement and choose the type of interaction:

- In TC-1, the weak chemical insufficiently removes dirt.
- In TC-2 the strong chemical damages surface (harmfully).

To solve TC-1, when chemical is weak, we must improve the insufficient action weak chemical insufficiently removed dirt.

To solve TC-2, when chemical is strong, we must remove or nullify the harmful action strong chemical damages surface.

Step 2 Validate the interaction is direct between A and B.

Check there are no intermediate objects and functions that link between tool and objects (S2 and S1).

Actions are direct.

Step 3 Identify S2, S1, the field and re-state the type of interaction.

227

For TC-1
- Weak chemical insufficiently removes dirt via a chemical field.

For TC-2
- Strong chemical damages surface via a chemical field.

Step 4 Follow the flowchart.

Flowchart is a guideline. For additional ideas use the rest of the standards to prompt ideas.

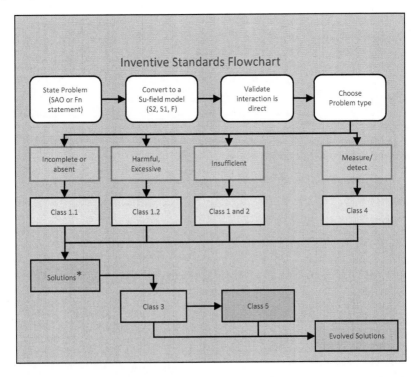

* Solutions are created by using Class 1.1, 1.2, 1, 2 and 4 to form an effective complete Su-field or an effective measurement and detection Su-field. The effective Su-field or "system" may be further developed by Class 3 and made more ideal by Class 5 inventive standards resulting in more evolved solutions.

Step 5 List ideas prompted by the inventive standards.

TC-1 Action is insufficient, so we first apply Class 1 and 2. TC-2 is harmful we apply Class 1.2

For TC-1 inventive standards 1.1.3 and 1.1.5 suggest ideas:

1.1.3 Introduce additives (S3) temporarily or permanently into S1 or S2, or between S1 and S2. (External Complex Su-field) – agitate the chemical with bubbles.

Standard 1.1.5 Change or modify the environment (may use additives, fields, parameters etc.). (Note this does not have a standard basic Su-field diagram to help explain). Heat the chemical or surface.

For TC-2 inventive standard 1.2.3 prompts an idea:

1.2.3 Introduce a sacrificial substance S3 to absorb or "draw off" the effect of a harmful action – add another chemical that forms a passive surface on the surface to stop the chemical attack.

We can continue creating many ideas including more "evolved" solutions from Classes 3 and 5 using the 76 inventive standards. Note the concepts using Su-field modeling and inventive standards to solve a technical contradiction are more specific than those created by the 40 inventive principles.

2. Advanced Tool 2: Solutions in Time and Space

We can define three separate time periods as we transition through a contradiction: T1 the time during TC-1, the time during State 1, T2 the time during the transition to State 2 and T3 the during TC-2 (State 2).

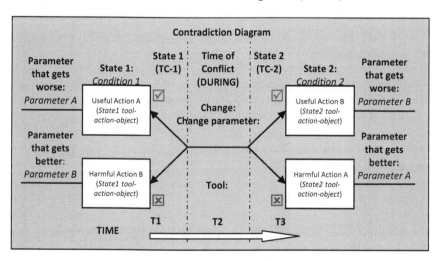

Note that T2, the time of conflict may be instantaneous. We illustrate it as a horizontal line rather than a point.

There is also a time before State 1 T- (before T1) and the time after State 2 T+ (after T3). It is important to consider solutions at each stage.

- T- the time in advance of State 1

- T1 the time during TC-1 (State 1)

- T2 the time during transition between states.

- T3 the time during TC-2 (State 2)

- T+ the time after State 2

During a contradiction, it is useful to illustrate the three states of the conflict at times T1, T2 and T3. **Drawing** is a powerful tool for releasing psychological inertia.

Note that standard ARIZ-85C only identifies three times as opposed to our five. ARIZ labels T2 the time prior to the conflict, T1 the time of the conflict and T3, the time after the conflict (sequence is T2, T1, T3). We will not use ARIZ conventions here; instead we have defined five time periods.

We use a six step algorithm for applying advanced Tool 2.

2.1 Solving the Technical Contradiction using Time and Space Resources Algorithm (Advanced Tool 2)

Step 1: State the two TC's.

Step 2: Complete the Contradiction Diagram.

Step 3: Draw the situation at T1, T2 and T3.

Step 4: Reference the drawing of T2, define: the Operating Time (OT) and Operating Space (OS).

Step 5: Re-state the two opposing requirements in T2.

Step 6: Ask the questions

- Is it possible to separate the opposing requirements in space?

- Is it possible to separate the opposing requirements in time?

- Can the problem be solved by doing "something" in advance (during T-)?

- Can the problem be solved by doing "something" after (during T+)?

In the same way we separated opposing **requirements** for solving a physical contradiction, we separate the opposing **actions** in time and space for a technical contradiction (we can also use the suggested inventive principles for how to perform the separation contained in the Table of Specific Inventive Principles to Solve Physical Contradictions see Appendix 7).

Examples

Example 1: The problem is a short bus can only carry a few passengers.

Step 1: State the two TC's

- TC-1: If the bus is short then it can easily turn corners but it holds few passengers.

- TC-2: If the bus is long then it holds many passengers but it cannot turn corners.

Step 2: Complete the Contradiction Diagram

(Exercise: identify parameters A and B and create ideas for solutions using the contradictions matrix. Parameters A and B reflect the actions).

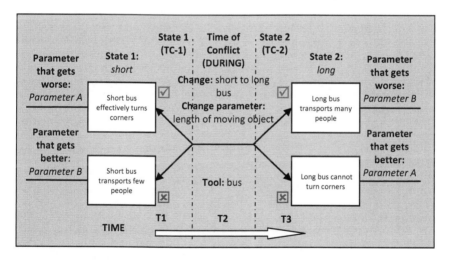

Step 3: Draw the situation at T1, T2 and T3.

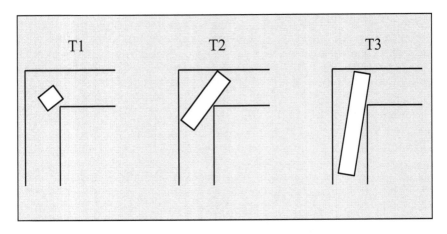

This is a top down view of the bus turning a corner. Draw side and any other views where applicable.

- T2 is the time of transition of the contradiction. A good way to think of it is as **the time of maximum opposition**. The bus is as large as it can be to hold maximum passengers but just can turn the corner.
- T1 describes the problem conflict in State 1. The bus is short, turning is not a problem, but there are few passengers.
- T3 describes the problem conflict in State 2. The bus is long, holds many passengers, but can't turn a corner at all.

We can try to find solutions in T1 and T3 as we have already discussed. *T2 illustrates our dilemma.* T2 defines when we have a conflict and where we have a conflict.

Step 4: Reference the drawing of T2, define: the Operating Time (OT) and Operating Space (OS).

o   The Operating Space (also known as the Z.o.C - The Zone of Conflict) is the region of space where the problem occurs.

o   The Operating Time (also known as the T.o.C - The Time of Conflict) is the time when the problem occurs.

In the bus example, the Operating Time is when the bus is turning a corner. The Operating Space is where the bus contacts the sidewalk.

Step 5: Re-state the two opposing requirements in T2.

The two opposing requirements during T2 are:
- To carry many passengers
- To turn corners.

Step 6: Ask the questions

- Is it possible to separate the opposing requirements in space?
- Is it possible to separate the opposing requirements in time?
- Can the problem be solved by doing "something" in advance (during T-)?
- Can the problem be solved by doing "something" after (during T+)?

In the bus example:

- Can the bus turn the corner in a different place from where it carries passengers?

No, the place the bus needs to turn corners is the same space as where it needs to carry many passengers.

We cannot separate opposing requirements in space.

- Can the bus turn the corner at a different time than when it carries passengers?

No, the time the bus needs to carry people is the same as when it needs to turn the corner.

- Can the problem be solved by doing "something" in advance (during T-)?

We could widen the corner, knock down some buildings.

- Can the problem be solved by doing "something" after (during T+)?

No, there is no T+ solution in this case.

The purpose of the first two questions is to prompt us to think of how to solve a problem by separating the technical contradictions opposing actions in time and space. The third and fourth questions focus on trying to solve a problem in advance (during T-) or after the contradiction (during T+). Let's try one we can separate in space.

Example 2: The problem is my coffee is too hot to hold.

Step 1: State the two TC's

- TC-1 If my coffee is hot then it is good to drink but it is difficult to hold.
- TC-2 If my coffee is not hot then it is easy to hold but it is not good to drink.

233

Step 2: Draw the Contradiction Diagram

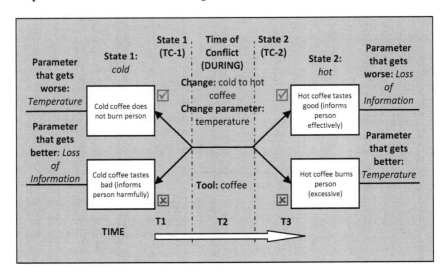

Step 3: Draw the situation at T1, T2 and T3.

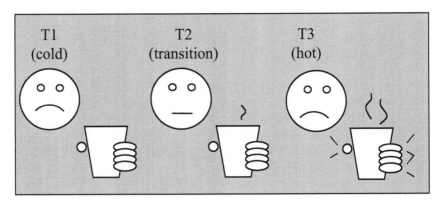

In T2 the coffee is as hot as can be held to maximize tasting as good as it can (maximum compromise).

Step 4: Reference the drawing of T2, define: the Operating Time (OT) and Operating Space (OS).

The Operating Space is the space where the coffee needs to taste good and the space where it needs to be not hot (separate locations).
The Operating Time is when the coffee is being drunk and the coffee is being held (separate times)

Step 5: Re-state the two opposing requirements in T2.

234

The two opposing requirements during T2 are
- To be held without burning (to be cool).
- To taste good.

Step 6 ask the questions

- Is it possible to separate the opposing requirements in space?

- Is it possible to separate the opposing requirements in time?

- Can the problem be solved by doing "something" in advance (during T-)?

- Can the problem be solved by doing "something" after (during T+)?

- Can the place where the coffee has to be cool be separated from where the coffee must taste good?

Yes, the place coffee must taste good is inside the cup, the place it must not burn is outside the cup (note that here "burn" means to heat painfully, not to literally ignite and burn). We can separate them by using a glove, cup holder, second cup etc., to insulate the hand from the hot cup. So we can have the hot coffee without compromise.

- Can the time when the coffee has to be cool be separated from when the coffee must taste good?

Yes, the coffee could be cool in the cup but hot to taste good. We could try cold coffee and heat it on its way from the cup to the mouth (a straw that heats?).

- Can the problem be solved by doing "something" in advance (during T-)?
Use thermally insulated cup to begin with.

- Can the problem be solved by doing "something" after (during T+)?
Put some ice in your hand to recover from the hot sensation.

We created our own ideas for solutions. We could also search for scientific effects by defining the function statement to be searched for (how to heat water, etc). There are also the specific inventive principles that can be used for prompting ideas for how to separate in time and space. They are used for solving a physical contradiction, but apply equally to separating a technical contradiction's conflicting requirements in time and space. The Table of Specific Inventive Principles to Solve Physical Contradictions can be found in the Appendix 7.

3. Advanced Tool 3: Use Smart Little People to solve a Technical Contradiction

Smart Little People is a tool for trying to find a solution using creative imagination by describing the problem in terms of little people. It is a tool to release psychological inertia. It is a similar tool to "Synectics" where the user imagines him or herself inside or as part of the problem. It has an advantage over imagining yourself inside the problem because you can separate the "little men" into many little men, it is difficult to think of yourself as many little men.

Imagine the useful and harmful action in a technical conflict as being performed by at the micro level by Smart Little People, able to take action, possessing the necessary special powers to perform the actions, in infinite numbers and groups (A, B, C, etc). The little men can be part of the system already (molecules performing a function) or can be added to provide an action to perform the useful or eliminate the harmful function.

First draw the contradiction with groups of little men performing what they are currently doing (the good and the bad actions). Then draw what the little men "should be" doing to solve the problem. From the "should be" try to from practical solutions

Take the example:

o   TC-1 If chemical is weak then chemical does not damage the surface but weak chemical insufficiently removes dirt.
o   TC-2 If chemical is strong then chemical effectively removes dirt but strong chemical damages the surface.

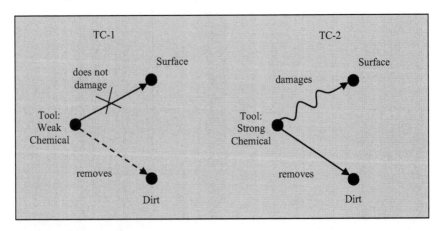

For TC-1, we cannot draw the activity of a null positive action, "weak chemical does not damage surface." We can only draw the action of little men insufficiently removing dirt. Both the harmful and useful action in TC-2 can be drawn; we can draw good little men removing dirt and bad little men damaging the surface. Let's draw the two situations.

TC-1 (T1)   TC-2 (T3)

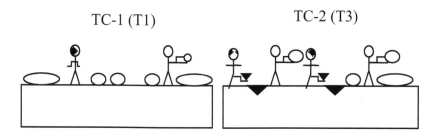

In TC-1 there are little men insufficiently removing dirt and no-one is damaging the surface.
In TC-2 there are little men effectively removing dirt, there are bad little men damaging the surface.

We should draw the contradiction at all three parts of the contradiction during T1, T2 and T3 to effectively use Smart Little People not just T1 and T3.

T2 illustrates the dilemma.

T2

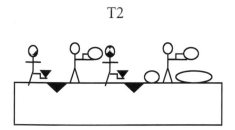

From the drawings, are there ideas inspired by what little men **should be** doing?
- Change the surface so the little men can't damage it. Add a protective coating.
- Add another group of little men that fill in the damaged surface.

The idea prompted is to add a chemical that is attracted to the areas that are exposed.

Because the problem is different in TC-1 to TC-2, draw both TC-1 and TC-2 situations and *draw the situation during the transition T2 when the opposition*

237

is at its greatest; this helps us think about solutions that might occur during T1, T3 and T2 during the time of conflict - dynamic solutions that may provide ideas for how to solve the conditions during the transition (something that appears only when needed at the transition).

3.1 Algorithm for SLP

> **Step 1**. State the technical contradiction as two technical conflicts; TC-1 and TC-2. Complete the contradiction analysis (use the Contradiction Diagram).
> **Step 2**. Draw the situation at T1, T2 and T3.
> **Step 3**. From the drawings, identify the "should be." What should the little men be doing? Assume you can add as many groups with special powers, etc. to solve the problem in TC-1 (T1), during the conflict (T2) or in TC-2 (T3). Create several "should be" scenarios for each problem stage, if needed, to make as many ideas as possible.
> **Step 4.** Identify practical ideas/solutions based on the various "should be" solutions.

4. Substance-Field Resource Analysis (required for Tools 4 and 5)

To solve the contradiction, we could choose to introduce new resources, new materials, objects, additional parts, etc. We could allow maximum changes to the system. To solve the problem more ideally, we should use existing available resources. Find a simple solution that does not make the system more complex, add more cost or have harmful consequences.

$$Ideality \sim \frac{Functionality}{Cost + Harmful\ Effects} \sim \frac{Functionality}{Expense} \sim Value$$

If we first try to find a solution that requires minimal changes to the system, minimal introduction of new resources we will identify the most ideal solution. If new resources are needed then they should be revealed by exhausting ideas that use available resources first.

To solve a contradiction we make a list of all available resources. This list will be used for Tools 4 and 5. The list includes the substance and field resources of all objects of the system, nearby or free resources, even cheap disposables that could be made available.

We can create an inventory of available resources in a table. From our analysis of a technical contradiction we form the contradiction into the tool and object(s).

The inventory is exactly the same for TC-1 and TC-2, the same resources are available.

4.1 Example of Substance Field Resource Analysis Table

1. System Resources (Internal) – resources of the tool and object(s) [product(s)] that make up the problem (the system).

   a) Tool
   b) Object 1
      Object 2
   A technical contradiction may have one or two objects but not more.

2. Supersystem Resources (External)

   a) SFR resources of the operating space environment at the problem.
   b) SFR resources common to any external environment (e.g. gravity, the earth's magnetic field, air, etc.).

3. Surrounding Resources

   a) Waste material from outside the system.
   b) Cheap or inexpensive material with negligible value.

We can put this in the form of a table:

Substance-Field Resource List			
Internal System Resources	Substance	Field	Parameter
1 System Resources			
Tool:			
Object 1:			
Object 2: (if present):			
External Resources			
2 Supersystem Resources			
Operating Space Environment			
Common to any Environment			
3 Surrounding Resources			
By-products, waste material			
Nearby systems			
Cheap or free available resources			

4.2 Example of SRF list creation

Let's make an SFR list for the example problem of a nail bending when hit with a hammer.

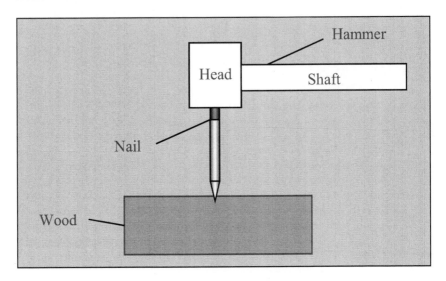

- ○ TC-1 If the hammer force is low then the nail does not bend but the nail moves slowly.
- ○ TC-2 If the hammer force is high nail moves quickly but the nail bends.

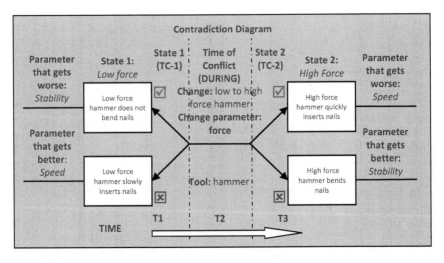

Referring to the Graphical Representation and Su-fields helps clarify the tool and object(s).

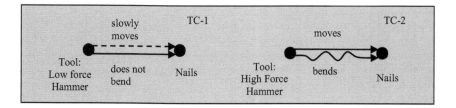

Tool: Hammer, Object 1: Nails, Object 2: None

Now make an inventory of all the substances, fields, and their parameters. This list of "raw" resources can be fairly extensive. The level of detail can be extended if the initial list does not provide good solutions. We will also consider combinations, derivatives, spaces, etc, as secondary resources in addition to these "raw" resources if a solution is not found using the raw resources.

SFR Resources 1 – raw resources
SFR List

Substance-Field Resource List			
Internal System Resources	Substance (thing)	Field	Parameter
1 System Resources			
Tool: hammer	Hammer head Hammer shaft	Mechanical force	Head: strength (of force),shape, hardness, internal composition , temperature, viscosity, uniformity, internal structure, weight, color, motion, transparency Shaft: Shape, flexibility, hardness, etc.
Object 1: nail	Nail	Speed of nail	Material, shape, length, internal structure, sharpness, temperature, uniformity, elasticity, thickness, roughness, color, porosity, internal structure, chemical properties, magnetic properties, direction, motion, etc.
Product 2: (if present): N/A			
External Resources			
2 Supersystem Resources			
Operating Space Environment	Wood Person		Wood: Density , hardness, internal tension, thickness, porosity, shape,

241

			structure, color, temperature Person: strength, weight, size.
Common to any Environment	Air	Gravity Earth's magnetic field	Atmospheric pressure, ambient temperature, ambient light.
3 Surrounding Resources			
By-products, Waste material	Discarded nails	Heat generated, Sound of hammer blow	Quantity of heat, Sound level, frequency, duration of sound.
Cheap or free available resources	Water supply is nearby		

We make an "inventory" of all the SFR resources.

- A substance is any object or material present that is not a "field."
- A field is as we define it in Su-field modeling includes all the strict definitions used for fields in science (weight, pressure, temperature, magnetic properties, etc.). But we also include a wider type of definition: such as taste, pressure field, smell (olfactory) field, stickiness field, biological, chemical, acoustical, optical field, etc
- A parameter is a property or feature or characteristic of a substance or field can be quantified or measured.

If the above "raw" resources do not provide a solution, try the following:

4.3 Secondary Resources

Resources 2.1 Use a mixture of resources.
Earlier we tried single resources in the SFR list. Now you can try combined (as many as you like).

Use a mixture of resources	

Resources 2.2 Consider using empty space or a combination of a substance with empty space to solve the problem (empty space can mean space, gap, cavities, bubbles etc).

Use empty space (gaps, bubbles, cavities, voids etc.).	

Resources 2.3 Consider using derivatives (or a mixture of derivatives with empty space)

242

A derivative is a changed resource, change **phase**, parts of materials, burnt remains, decomposed materials, molecules etc anything we can derive from an existing resource or resources.

Use derived resources	
Use derived resources with empty space **(gaps, bubbles, cavities, voids etc.).**	

Resources 2.4 Hidden resources.

These are resources that are present but often assumed to not be present or are not known to be present but actually are. For example: components of air, contamination level in water, foreign materials (wood also contains moisture) etc.

Use hidden resources	
Use contaminants, substances that may form be new mechanisms, etc.	

The SFR list is not a tool for solving a problem; it is an inventory of resources that will be used to solve the problem using tools 4 and 5.

5. Advanced Tool 4: The Ideal Final Result 1 (IFR-1)

5.1 State the Technical Contradiction in the form of the Ideal Final Result 1

IFR-1 is the solution statement: X-resource itself provides <useful action> and eliminates the <the negative action> without making the system more complex and without additional harmful consequences.

To form IFR-1 (for TC-1 and TC-2) we first create an X-component statement for each technical conflict using the useful and harmful actions we identified from the analysis of the contradiction (Contradiction Diagram).

Example
o   TC-1 If the hammer force is low then the nail does not bend but the nail moves slowly.
o   TC-2 If the hammer force is high then the nail moves quickly but the nail bends.

For each of these conflicts we can create an imaginary solution statement of the form "something," an "X-resource" that solves our conflict. For example, we could make a statement in the form:

- X-resource provides (or preserves or maintains) <the useful action> and eliminates <the harmful action>.

It is a statement of the final result we want to happen. At first we don't know what that "something" is or how it solves the problem but we can make the leading statement. This is a "solutions first" approach to solving a problem. It is very powerful for releasing psychological inertia.

Take the following simple example we showed in Chapter 1: in the illustration below, which string should the mouse pull on to fetch the cheese, A, B or C?

The simplest way to solve this problem is to start from the cheese and follow the string back to identify string A, B or C. By identifying the final result first, we can more easily solve the problem.

In our example of the hammer, we can state the two technical conflicts TC-1 and TC-2 as:

- TC-1 (weak force): X-resource provides no bending of the nail and eliminates slow speed of insertion.
- TC-2 (strong force): X-resource provides fast nail insertion and eliminates the bending of the nail.

Now let's state this solution more ideally. Ideality requires minimal cost or harmful effects with maximum benefits (functionality). We would like to solve the problem with minimal changes and minimal introduction of resources. We want X-resource ideally to solve the problem by *itself*.

- TC-1 (weak force): X-resource *itself* provides no bending of the nail and eliminates slow speed of insertion.
- TC-2 (strong force): X resource *itself* provides fast nail insertion and eliminates the bending of the nail.

244

We are not choosing to use medium strength force (i.e. a trade-off or compromise). We choose weak or strong force. Our target solutions are:

- When the force is weak, X-resource itself provides no bending and eliminates slow insertion.
- When the force is strong the X-resource itself provides fast insertion and eliminates bending of the nail.

Our solution when forming the ideal final result statements, the IFR-1's, is not targeted at solving the problem in advance or performing a recovery solution after but to eliminate the problem during the conflict and to eliminate the problem where and when it occurs.

We create two IFR-1 statements, one for each TC of the form:

- X-resource **itself** provides <the useful action> and eliminates the <the harmful action> without making the system more complex and without additional harmful consequences.

Note that sometimes the words during the operating time in the operating space are added.

- X-resource **itself** provides <useful action> and eliminates the <the negative action> during the operating time in the operating space without making the system more complex and without additional harmful consequences.

We will not use them. They make the statement overcomplicated and confuse the user who tries to think only of ideas that occur during TC-1 or TC-2 and not during the transition. The operating time is in fact T1 and T2 and T3 as we defined earlier for tool 2.

In our example, the two IFR-1s are:

- TC-1 (weak force) X-resource itself provides no bending and eliminates low insertion speed without making the system more complex and without any harmful consequences.

- TC-2 (strong force) X-resource itself provides high insertion speed and eliminates bending of the nail without making the system more complex and without any harmful consequences.

To aid thinking and release psychological inertia, users should *exaggerate* the actions. For example:

- TC-1 (weak force) X-resource itself provides absolutely zero bending and eliminates slow insertion speed without making the system more complex and without any harmful consequences.

- TC-2 (strong force) X-resource itself provides extremely high insertion speed and completely eliminates any bending of the nail without making the system more complex and without any harmful consequences.

You may pursue one or both TCs. The one in which the main useful function is better delivered is normally recommended. In this case the main useful function is to insert a nail, which is best performed when the force is strong (TC-2). Choosing the TC that does not provide the main useful function best, as we discussed previously often leads to more radical ideas. We won't choose a specific technical conflict but will pursue both.

Now we form the IFR-1. IFR-1 is actually a list of statements where the "X-resource" part of the sentence we constructed is substituted by each of the resources we identified in the SFR list we created.

This step provides a significant release of psychological inertia for how to solve the technical contradiction. It is a "solutions first" approach to finding a solution.

IFR-1 List for TC-1 (weak force) – we will use the exaggerated version.

The **hammer head** itself, provides absolutely zero bending and eliminates slow insertion speed without making the system more complex and without any harmful consequences.
The **hammer shaft** itself, provides absolutely zero bending and eliminates slow insertion speed making the system more complex and without any harmful consequences.
The **mechanical force** itself, provides absolutely zero bending and eliminates slow insertion without making the system more complex and without any harmful consequences.
The **shape of the hammer head** itself, provides absolutely zero bending and eliminates slow insertion speed without making the system more complex and without any harmful consequences.
The **internal composition of the hammer head** itself, provides absolutely zero bending and eliminates slow insertion speed without making the system more complex and without any harmful consequences.

We continue through all of the resources, collecting ideas.

Ideas created are: for a weak force, we need faster inserting with no deterioration in bending,

246

1. The **porosity of the wood**. Pre-drill holes where they are to be inserted so the nails don't bend but insert faster.

2. The **internal tension of the wood**. Stress the wood so it is easier to insert a nail (less resistance to the nail).

And we can create many more as we go through the list prompted by the Ideal Final Result statements,

Perform the same for TC-2.
IFR-1 List for TC-2 (strong force)

X-resource itself provides extremely high insertion speed and completely eliminates any bending of the nail without making the system more complex and without any harmful consequences.

The **hammer head** itself, provides extremely high insertion speed and completely eliminates any bending of the nail without making the system more complex and without any harmful consequences.
The **hammer shaft** itself, provides extremely high insertion speed and completely eliminates any bending of the nail when the hammer is driving in the nail where the nail is driven into the wood without making the system more complex and without any harmful consequences.
The **mechanical force** itself, provides extremely high insertion speed and completely eliminates any bending of the nail when the hammer is driving in the nail where the nail is driven into the wood without making the system more complex and without any harmful consequences.
The **shape of the hammer head** itself, provides extremely high insertion speed and completely eliminates any bending of the nail when the hammer is driving in the nail where the nail is driven into the wood without making the system more complex and without any harmful consequences.
The **internal composition of the hammer head** itself, provides extremely high insertion speed and completely eliminates any bending of the nail when the hammer is driving in the nail where the nail is driven into the wood without making the system more complex and without any harmful consequences.

We continue through all of the resources, collecting ideas.

Ideas created are: for a strong force, we need no bending with no deterioration in speed.

1. The **porosity of the hammer head itself.** Use moveable pins that recess into the head, the pins stop bending (lateral movement) and allow high force.

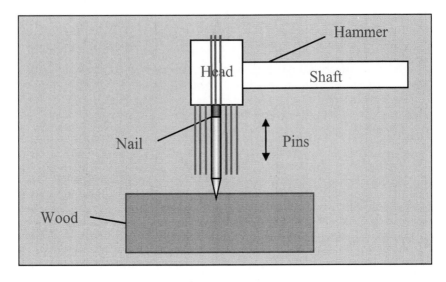

This idea prompts the idea of a "nail gun." The nail is "fired" down a tube

2. The **thickness of the nail itself**: Use two nails instead of one.

We continue through all of the resources, collecting ideas.

If the "raw" resources do not provide a solution, apply the secondary resources

Secondary Resources
Resources 2.1 Use a mixture of resources.
Resources 2.2 Consider using empty space or a combination of a substance with empty space to solve the problem (empty space can mean space, gap, cavities, bubbles. Etc.).
Resources 2.3 Consider using derivatives (or a mixture of derivatives with empty space).
Resources 2.4 Hidden resources

IFR-1 during T2 (transition)
If IFR-1 for TC-1 and TC-2 do not provide solutions even with secondary materials, then consider whether the resources can solve the problem during the transition (during T2). Form the IFR-1

- X-resource eliminates the harmful action and provides the useful action during time T2.

The useful and harmful actions are obtained from **the two opposing requirements in T2**.

5.2 Algorithm for IFR-1

Step 1. State the technical contradiction.
Step 2. Complete contradiction analysis and state TC-1 and TC-2 in terms of opposing actions. Identify the tool and object(s)
Step 3. Create a Substance-Field Resource list (see SFR table).
Step 4. For TC-1 and TC-2 create the IFR-1 statement:
TC-1 (State 1): X-resource itself provides <useful action> and eliminates the <the negative action> without making the system more complex and without additional harmful consequences.
TC-2 (State 2): X-resource itself provides <useful action> and eliminates the <the negative action> without making the system more complex and without additional harmful consequences.
Step 5. Exaggerate the statements.
Step 6. Create the IFR-1 list by inserting each resource into the IFR-1 statement for both TC-1 and TC-2.
Step 7. List ideas.
Step 8. For more ideas use secondary resources: 1. Combinations of resources, 2. Use emptiness (gaps, bubbles, cavities, voids etc.) 3. Use derivatives (also combine derivatives with emptiness), 4. Hidden resources.
Step 9. Apply IFR-1 during T2. Consider if each resource can solve the problem during the transition between states (T2).

Advanced Tools for Solving Physical Contradictions

6. Advanced Tool 5: Solve the Physical Contradiction by SFR Resource (IFR-2)

6.1 Convert the Technical Contradiction to a Physical Contradiction (IFR-2)

A physical contradiction can be stated: "X-resource (or part of it) itself, should be A and anti-A."

A physical contradiction is formed when the opposite property (parameter, characteristic or feature) is needed for an object, system, or part of an object. This is an X-resource. Opposite properties are "A" and "anti-A" which form antonym pairs: hot-cold, present-absent, big-small, blue- non-blue, solid-liquid, etc.

We can form a physical contradiction when:
- X-resource needs to have one property "A" to perform one action but the opposite anti-A to eliminate the harmful action.

- X-resource needs to have one property "A" but doesn't have that property or has the opposite property.

For example:

- The nail should be narrow to be inserted quickly but wide to not bend (narrow and wide).
- The wood should be non-solid to allow the nail to be inserted quickly but it should be solid because it is wood (non-solid and solid).

To identify opposing properties, it is useful to first think what parameter the useful action is a function of, similar to how we formed physical contradictions for the "Physical Contradiction Algorithm using Basic Tools." For example, an umbrella must stop rain and be easily carried. The useful action - to stop rain, is a function of size. If size is big then the action to stop rain improves. The antonym of large is small. If being small supports the elimination of the harmful action then a physical contradiction is formed. The harmful action is to not be easily carried. The umbrella should be large and small.

The key difference between the basic tool for forming a physical contradiction and the advanced tool is that we use the Substance-Field Resource list. Each resource may produce several physical contradiction problems so we create many more physical contradiction statements in contrast to when we have no Substance-Field Resource list.

We use an IFR-2 table to compile contradiction statements. For each resource we try to create macro and micro level physical contradiction statements.

Referring to the SFR table we created for the hammer and nail problem, first complete the column for macro level physical contradictions for each resource

Note that it is useful to exaggerate the physical contradiction as we did with the technical conflicts (the high force inserts the nail with tremendous speed, with absolutely no bending). Exaggerating the useful and harmful actions releases psychological inertia. In the IFR-2 table below, try to create antonym pairs for each resource.

IFR-2 Table (physical contradictions for each resource at the macro and micro level).

X-resource should be <A> to provide the exaggerated useful action and <anti-A> to eliminate < the exaggerated harmful action>.

Let's brainstorm opposite physical requirements for each resource (substance, field and parameter) at the macro level first. So complete only column 2, we will compile column 3 after column 2 is completed.

X-resource should be <A> to insert the nail with tremendous speed and <anti-A> to eliminate all bending.

IFR-2 Table

Resource	Physical Contradiction (**macro** level)	Physical Contradiction (**micro** level)
Hammer head	**Force** should be **high** to insert quickly and **low** to not bend. **Shape** must fit around the nail as it is inserted to stop bending but not fit around the nail to allow the hammer to reach the wood. **Solid** to be inserted quickly, **non-solid** to stop bending. **Motion** should be large to force in the nail small to stop bending. Etc.	
Hammer shaft	**Rigid** to insert the nail quickly but **flexible** to not bend the nail.	
Nail	**Narrow** to be inserted quickly and **wide** to not bend. **Short** to be inserted quickly and **long** because it's a nail and must secure something. Etc.	
Wood	**Soft** to insert quickly **hard** because it's wood.	
Air	**Solid** to stop bending, **non-solid** because it's gas.	
Etc		

After the table is complete for column 2, try to express each macro physical contradiction at the micro level column 3.

This is done by trying to state the macro physical contradiction in terms of moving particles, grains, atoms, molecules, etc. After this you could try to form a micro physical contradiction for each of the macro ones you have identified.

IFR-2 is the combined list of macro physical contradictions and micro physical contradictions.

251

Resource	Physical Contradiction (macro level)	Physical Contradiction (micro level)
Hammer head	**High force** to insert quickly, **low force** to not bend (the PPC). **Shape** must fit around the nail as it is inserted but not fit around the nail to allow the hammer to reach the wood. **Solid** to be inserted quickly, **non-solid** to stop bending. **Motion** should be large to force in the nail small to stop bending. Etc	Force: No ideas Shape: The hammerhead should be made of fluid particles to stop the nail bending and solid particles to drive in the nail. Solid: No ideas Motion: Motion should be many short micro-hammer blows to not bend and large to be inserted quickly
Hammer shaft	**Rigid** to insert the nail quickly but **flexible** to not bend the nail	No ideas
Nail	**Narrow** to be inserted quickly and **wide** to not bend. **Short** to be inserted quickly and **long** because it's a nail etc.	No ideas
Wood	**Soft** to insert quickly **hard** because it's wood.	No ideas
Air	**Solid** to stop bending, **non-solid** because it's gas.	No ideas
etc		

By thinking in terms of the micro level, particles, etc. we can sometimes release psychological inertia to create new ideas that were not inspired at the macro level. From "motion should be many short micro-hammer blows to not bend and large to be inserted quickly" we get the idea of using many small quick taps, creating the idea of a vibrating hammer.

We identified some physical contradictions. Let's try to solve them. For each contradiction formed at IFR-2 try to solve by satisfaction, bypass then separation. Let's try to solve the physical contradiction: **Wood must be soft to insert nail quickly and hard because it's wood.**

When the physical contradiction has been defined it is useful to use the following template for solving a physical contradiction by separation, satisfaction and bypass.

Satisfaction	Solution
Can the contradictory requirements be solved by satisfaction?	**Use an extremely hard thin nail (titanium).** **Use wood that is soft but then solidifies.**

Try to determine a method to satisfy the problem - search scientific effects, use Table of Specific Inventive Principles to Solve Physical Contradictions.

252

Bypass	Solution
Can the contradictory requirements be solved by bypass?	**Use a screw and electric screwdriver. Use glue, etc.**

Try to determine a method to bypass the problem - search scientific effects, use Table of Specific Inventive Principles to Solve Physical Contradictions.

Separation in Space	Where
Where is there a requirement to be < A >? **Soft**	**wood**
Where is there a requirement to be < anti-A >? **Hard**	**wood**

If there is a requirement to be A and anti-A at the same location, then they cannot be separated in space.

Here the requirements are in the same location therefore we cannot separate the contradictory demands in space

Separation in Time	When
When is there a requirement to be < A >? **Soft**	**When nail is being inserted**
When is there a requirement to be < anti-A >? **Hard**	**After nail has been inserted**

If there is a requirement to be A and anti-A at the same time, then they cannot be separated in time.

Here we *can* separate contradictory demands in time.

Separation in relation	For what or whom?
For what is there a requirement to be < A >? **soft**	**Inserting the nail**
For what is there a requirement to be < anti-A >? **hard**	**Holding the nail**

If there is a requirement to be A and anti-A for the same action or object, then they cannot be separated in relation.

It (the wood) must be hard and soft for the nail. We cannot separate contradictory demands in relation.

Separation at system level (parts v whole)	Level
For what level of the system is there a requirement to be < A >?	**Must be soft at system and parts level.**
For what level of the system is there a requirement to be <anti- A >?	**Must be hard at system and parts level.**

If there is a requirement to be A and anti-A at all different system levels, then they cannot be separated at system level.

We cannot separate contradictory demands at system level.

So we can separate the contradictory demands, that wood must be hard and soft in time.

We can try to create ideas of our own and we can refer to the Table of Specific Inventive Principles to Solve Physical Contradictions (Appendix 7).

From the table, the recommended inventive principles to use to separate in time are: 15 Dynamization, 10 Prior Action, 19 Periodic Action, 11 Beforehand cushioning, 16 Partial or excessive action, 21 Skipping, 26 Copying, 18 Mechanical Vibration, 37 Thermal Expansion, 34 Discarding and recovering, 9 Prior counter-action and 20 Continuity of useful action

Solutions/Ideas (add to solutions bank)

- Prior action: pre-drill holes so the nails insert quickly and don't bend.

- Mechanical vibration/periodic action: use lots of small blows quickly.

- Beforehand cushioning, surround the nail with a cushion that stops bending but collapses as the nail is inserted.

6.2 Algorithm for solving Physical Contradictions by Resource (IFR-2)

Step 1. Create the physical contradiction statement: X-resource (or part of it) should be "A" and "anti-A."

Where X-resource is a system, part of a system, or a component, "A" and "Anti-A" are an antonym pair (hot and cold, large and small, present, absent etc. of any property for X-resource).

- X-resource needs to have one property A to perform the useful action but the opposite anti-A to eliminate the harmful action.
- X-resource needs to have one property A but doesn't have that property.

Step 2. Create or refer to the Substance-Field Resource List used for IFR-1.

Step 3. Try to create one or more physical contradictions for each resource. Consider exaggerating the actions.

Step 4. Solve each physical contradiction by satisfaction, bypass or separation.

- Consider solutions during T1, T2 and T3. Even though we form only one physical contradiction (see above section in Chapter 9 Part 1. **Physical Contradiction - One Physical Conflict**, the problem may be solved in State 1, State 2 or during the transition between states).

Use the template for solving a physical contradiction by separation, satisfaction and bypass.

Step 5. Try to form each physical contradiction at the **micro level** (sometimes releases psychological inertia) and solve using satisfaction, bypass, or separation.

Step 6. Try to form physical contradictions using the additional resources for combined resources, derived resources, emptiness hidden resources and solve using satisfaction, bypass, or separation.

Step 7. Search for **scientific effects** use the Algorithm for How to Search for Scientific Effects to Solve a Physical Contradiction (see Chapter 7).
Simplified ARIZ

7. Simplified ARIZ the Algorithm for Solving Inventive Problems

We have combined the advanced methods above with the basic methods discussed in Chapter 9 Part 1 into a single algorithm for solving contradictions. It is a simplified form of ARIZ. It contains four main phases A, B, C and D. The algorithm describes the process for applying the Contradiction Problem Solving Flowchart - Simplified ARIZ shown in Chapter 9 Part 1 Section 2.5 and Appendix 87.. Details of Simplified ARIZ are provided in the **Appendix 8.8 – Simplified ARIZ Algorithm**

Simplified ARIZ (18 Steps)

A. Form and Solve Technical Contradiction using Basic Tools (10)
B. Form and Solve the Physical Contradiction using Basic Tools (3)
C. Form and Solve the Technical Contradiction using Advanced Tools (4)
D. Form and Solve the Physical Contradiction using Advanced Tools (1)

The number of steps in each section ABC and D are shown in parenthesis; the steps include running multi-step algorithms. The "official ARIZ" version ARIZ-85C which many users difficult to follow is also discussed in detail in Appendix 8.9.

Chapter 10

Subversion Analysis

Chapter Contents

1. Introduction to Subversion Analysis
  - o SA-1 Root Cause Analysis
  - o SA-2 Failure Prevention

2. Subversion Analysis-2 (Failure Prevention)
  2.1 Algorithm for Subversion Analysis 2

Chapter Summary: *There are two types of subversion analysis.*

Subversion Analysis 1 (SA-1) is used for root cause analysis; we brainstorm ideas to determine why a specific problem occurred by identifying ways to cause it. We discuss SA-1 in Chapter 11 – Root Cause Analysis Incorporating TRIZ Tools.

Subversion Analysis 2 (SA-2) is a tool used to identify failure modes of a technical system or process. We use CEC-4, with the starting problem defined as "the system failed" to create a list of what can go wrong and try to put measures in place to prevent them. A plan is developed to implement solutions ahead of time to prevent problems. We use the word "subversive" because we try to solve problems by causing them.

1. Introduction to Subversion Analysis

There are two types of subversion analysis:

- SA-1 To assist root cause analysis
- SA-2 Failure Prevention

Subversion Analysis (SA) is a method of trying to reveal root cause or predict a future potential problem by subversion. To be subversive is to try to undermine or sabotage the situation. By trying to cause a problem or failure in a technical system or process we can try to:

- SA-1: reveal the root cause of a (reactive) problem by trying to cause it.
- SA-2: try to identify the ways a system might fail in future.

If we know what failures might occur in the future we can solve them proactively and design solutions that prevent failures into our system or process.

In this chapter we will discuss only SA-2. SA-2 is an analytical tool for preventing future failures; these are Type 4 problems in our TRIZICS Roadmap. SA-1 is for a Type 1 problem, it is one of the tools we use to reveal root cause and will therefore be included in Chapter 11.

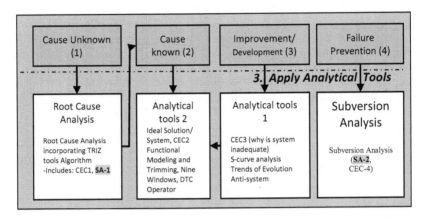

Unfortunately many problem solvers, engineers, etc. will apply SA-1 much more than SA-2, trying to find the cause of failures rather than preventing them in the first place. Many engineers and problem solvers see themselves as problem solvers reacting to problems rather than problem preventers proactively eliminating failure modes. A more ideal problem solver works on failure prevention.

2. Subversion Analysis – 2 (SA-2 Failure Prevention)

The objective is to try to identify what can go wrong with a technical system or technical process then create solutions and preventative measures to eliminate or mitigate the failure mode. Try to focus on the most likely failures that can be created by the available resources. De-prioritize unlikely external possibilities – such as, a meteorite could hit.

2.1 Algorithm for Subversion Analysis 2 (Failure prevention)

Step 1 Identify the system
- Identify the system or process or part of the system or process to be evaluated.

Step 2 Collect Information

258

- Collect all technical information including: pictures and drawings, etc.
- Describe how the system operates. It may be useful to create a basic functional model to gather data about the operation and quality of interactions.
- List key system components. It may be useful to create an SFR-list (tool, product, internal and external resources including the environment – substances, fields and their parameters).
- For all systems, procedures, parts and materials describe:
  - how quality is maintained and controlled including how they are stored and transported.
  - how parts, resources, etc. change, erode or deteriorate over time.

Step 3 List what can go wrong (Make a list of failure modes using CEC-4)

- Perform a Cause-Effect Chain analysis (CEC-4) with the problem statement as "**the system failed**." Use the information gathered in Step 2 to help prompt ideas as the chain is being built. The chain prompts us to think of ways to cause problems (subversive).
- Several chains should be made if many fail modes are identified and it is difficult to fit them all into one diagram.
- The CEC-4 diagram provides a brainstormed list of fail modes and the underlying reasons for why those modes might occur.

For example:

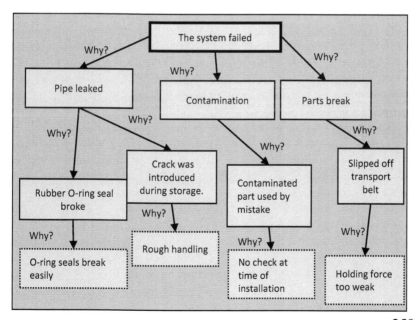

- After the chain is completed the user creates a list of all failure modes and the underlying causes. For each failure mode there may be several model scenarios for how the failure could occur.

- It may be useful to rank (prioritize) the identified potential failure modes and model scenarios in order of likelihood and problem severity, then prioritize which problems to pursue first based on the product of both rankings.

Failure Mode	Model Scenario (cause)	Solution/Prevention
1. The pipe leaks	O-rings are easily broken because they are fragile Cracks are introduced due to rough handling during storage.	
2. Contamination	Parts not checked for contamination at time of installation.	
3. Parts break	Force that holds them to the transport belt is too weak.	

Step 4: Create and implement solutions to prevent failure modes.

Failure Mode	Model Scenario (cause)	Solution/Prevention
1. The pipe leaks	O-rings are easily broken because they are fragile. Cracks are introduced due to rough handling during storage.	Find a source of o-rings that are not fragile. Implement new handling procedures and soft handling materials for storage.
2. Contamination	Parts not checked for contamination at time of installation.	Create an inspection test for contamination at install.
3. Parts break	Force that holds them to the transport belt is too weak.	Create a method for holding parts securely.

Often a standard in-the-box solution can be applied or if needed apply the TRIZ solutions tools to create breakthrough ideas.

Chapter 11

Root Cause Analysis - Incorporating TRIZ Tools

Chapter Contents

Chapter Summary: *Root cause analysis is first performed when we apply Standard Structured Problem Solving (see Chapter 2). Cause-Effect Chain, Fishbone Diagram and brainstorming tools are those conventionally used to try to find root cause. TRIZ was not originally developed to find root cause but several of the TRIZ tools (and some non-TRIZ tools) have been adapted to*

261

*form powerful new thinking tools to help find root cause and are used to create an augmented structured problem solving process for problems when root cause is difficult to identify. This alternative process is called "**Standard Structured Problem Solving Incorporating TRIZ tools for Root Cause Analysis**" and it should be used for Type 1 problems when root cause is elusive. The algorithm has an associated flowchart and template are included in the Appendices 8.5 and 8.6.*

1. Introduction

Classical TRIZ tools were developed to create innovative new ideas and breakthrough solutions, not to determine root cause. Classical TRIZ tools are used when the cause of a problem is known.

For example, I live in a hard water area, after a few weeks my electric kettle builds up a calcium residue. How do I remove and stop the formation of the residue? In this example we understand the root cause of the problem; the calcium is in the water supply and precipitates onto the walls of the kettle. To solve this problem, we can formulate it in any of the classical TRIZ formats (function statement, Su-field model, technical or physical contradiction) and create innovative ideas.

Now consider the more common situation in which cause is not known, for example, you are an engineer in a manufacturing plant, you discover the flow from one of a group of ten identical water tanks has fallen by 10%. Your task is to eliminate the drop in flow and stop it happening again. Root cause analysis is needed, if the cause of the problem is not identified, then you cannot solve it.

After investigation you discover there is a chalk build-up on the inside of the water tank, it has reduced water flow by 10%. What caused the chalk build up? The water supplied to each tank is from the same source. Root cause analysis investigations reveal many potential causes, but eventually the problem is traced to a wrongly calibrated valve that is used to provide acid to reduce the calcium build up.

Much of problem solving requires this type of root cause identification first rather than creating novel solutions for "understood" problems. This is a frequent source of confusion amongst engineers. The title "Inventive Problem Solving" is wrongly interpreted to include troubleshooting and root cause analysis which are of course problems that require inventiveness, but are not included in TRIZ. Much of "engineering problem solving" is basically troubleshooting to determine the root cause of a failure or unwanted result. TRIZ tools are often not needed because routine in-the-box thinking can frequently provide a robust fix when root cause is known and "inventive solution finding" (TRIZ) is not needed.

262

The TRIZICS roadmap starts with Standard Structured Problem Solving to ensure the problem is correctly and clearly defined and that root cause is known. The standard tools for finding root cause are brainstorming, Fishbone Diagram or Cause-Effect Chain, but even those may fail to lead us to find the cause. Let's discuss how to leverage the creative power of TRIZ tools by adjusting them to assist our thinking for root cause analysis - engineers can use TRIZ tools to help find root cause, with some minor adjustments.

What is root cause?

It is important to distinguish between the symptoms of a problem and the underlying cause(s) of a problem. If I cut my finger while slicing tomatoes with a sharp knife, I can stop the bleeding by using a "band-aid." This will solve the problem in the short term, if I get another cut I can add another "band-aid", and I can continue until I am covered in band-aids which creates secondary problems – it's inefficient to live with lots of "band-aids." In general it is better to solve the root cause of a problem rather than alleviate the symptoms. We should investigate what caused the cuts to prevent it happening again. A solution that addresses the symptoms should be regarded as a containment fix but not a root cause solution. Instead of addressing what happens, address why it happens.

Note that for any problem that needs root cause analysis, there may be a single root cause or there may be several root causes. For the example above these may be: the knife is too sharp, the accuracy of slicing is inadequate, the toughness of the skin is insufficient, the person forgot to wear gloves, forgot to wear their eyeglasses, etc. There may be several factors that combine to form root cause, for example, the cut may only occur when the knife is too sharp *and* I do not wear gloves.

Failures may also be intermittent, for example defects on manufactured parts may occur only on a Tuesday when the temperature is below 60F, the clean at a previous step was performed by water from tank number six and the parts have been in storage for more than 8 hours. To remove the problem at a root cause level, each single cause or at least one of the combination of causes should be stopped to eliminate the problem.

To find root cause we need to have sufficient information to analyze in order to create models of cause mechanisms. This normally requires evaluating routinely collected data, collecting additional new data and designing and performing experiments. Data collection and experimentation should be designed to reveal the circumstances in which the problem occurs and any other information supportive of defining a root cause mechanism. Given this information we can create root cause models that are consistent with the data (model scenarios) and design validation tests to confirm them as the cause.

The target of root cause analysis is to create a verifiable mechanism or model that describes how the problem occurs that is consistent with the data.

Sometimes verifying root cause is not within our capability or scope of work, we may not have the measurement tools or lab conditions needed to investigate the mechanism. However, we may be able to eliminate a problem without a verifiable model. For example, defective parts are made only when tool number three is used at step five and tool number seven is used at step twenty of the manufacturing process. All other tool combinations are not defective. After intensive investigation we find no measureable difference or circumstance related to these tools that could cause the problem. To solve the problem, we simply remove that particular combination. But did we eliminate root cause?

We did eliminate the cause, but there is no verifiable mechanism. Without a mechanism we cannot be sure the root cause is eliminated, for example any other combination of tools may start exhibiting the same problem as tool three and seven. The fix is a "band-aid" that addresses the symptoms.

We will not discuss how to collect and analyze data or design experiments; this is circumstantial and is well documented in engineering books for experimenters and data analysts. Frequently, despite best efforts and talented brains, after brainstorming, Fishbone analysis and model scenario building with expert data analysis, the root cause still cannot be found. It is for these situations that the rest of this chapter is written.

To begin, first perform the initial steps discussed in the Standard Structured Problem Solving roadmap discussed in Chapter 2.

3. Standard Structured Problem Solving *(see Chapter 2).*

Step 1: Define Problem
Step 2: Current Situation
Step 3: Analysis
       Data analysis and results
       Root Cause
              Brainstorming
              Create models
              CEC analysis
Step 4 Solutions
Step 5 Implementation

Root cause analysis is completed in Steps 1-3. But what if we still can't find root cause? How do we leverage TRIZ tools and release psychological inertia that will help identify it?

Before applying TRIZ tools, let's discuss and address some of the most common reasons that we fail to deliver a root cause.

4. Common issues that impede root cause determination

Psychological inertia, erroneous data, wrong assumptions and false conclusions are often why root cause is not identified. It is important to keep an open mind and eliminate the causes of psychological inertia. It is important to validate and challenge all data and assumptions. It is necessary to prove the data being used is valid and to ensure the correct conclusions are drawn.

Below is a list of frequent reasons root cause is not identified and recommended actions the problem solver should take to address those issues.

4.1 Frequent reasons root cause is not identified:

- **Experience**: too much experience or "expertise" can lead to psychological inertia and drive thinking in a "trained" direction closing off new ideas.
- **Fixed thinking techniques**: repeating the same procedural steps and using the same methods can lead to repeating the same result, creating the same ideas.
- **Group think**: over time, a set of individuals working on a project will tend to think the same way, believe the same conclusions and results. This group mindset leads to psychological inertia, as new members are introduced instead of pursuing new ideas provided by "a fresh pair of eyes." The group tries to assimilate new members to existing thinking.
- **Model worship:** a specific "favorite" model is pursued and alternatives are dropped.
- **False Information/incorrect data/false assumptions**: this is may be due to the way the data or information was collected. For example, incorrect calibration of a measurement standards, or simply incorrect information or facts have been obtained or assumed.
- **False Conclusion**: for example, the sun rises every day in the east. False conclusion - the sun revolves around the earth.
- **Hidden resources**: the problem is caused by contaminants, or secondary resources, etc. not recognized or which are overlooked by the problem solver.
- **Hidden mechanism**: mechanism may be a new or unusual phenomenon or be an effect outside the problem solvers field of engineering or science.
- **There is no actual problem**: the result is "normal" simply a "rarity" or "outlier" of the normal distribution.
- **There is more than one problem** (several different failure modes).
- **Insufficient technical knowledge**: this is rarely the reason for a problem's root cause not being identified. Normally such gaps in knowledge are quickly closed and problems solved.

4.2 Actions to address common issues that impede root cause determination

- Have new (different) people check all data and information to provide fresh thinking.
- Determine whether the conclusions can be wrong (be highly critical of all conclusions).
- Check the information is indisputable, assign a specific person (owner) responsible for checking the data
- Physically check and visually witness information or data rather than accepting validation from others.
- Always challenge calibration methods.
- Determine what potentially hidden or secondary resources might be present and how they could cause the problem.
- Describe a new or unusual mechanism that would have to exist to cause the problem.
- Demonstrate the problem is not simply an outlier (a rare but expected event and therefore not a "problem" at all).

Independent review/validation of each piece of data, assumption and conclusion is needed. It is useful to list all assumptions and conclusions and challenge each in turn. Re-checking the information, using different personnel is necessary.

If the information and assumptions relating to the problem have been validated, then we should apply adapted TRIZ tools to try to find root cause. But which tools do we choose and how do we use them?

5. Analysis of TRIZ tools – which of them can help determine root cause?

5.1 List of TRIZ tools

The tools we have defined in our TRIZICS Roadmap are:

Analytical Tools
- a) *S-curve analysis*
- b) *Trends of Evolution (also a solutions tool)*
- c) *Cause-Effect Chain Analysis*
- d) *Ideal Solution/System*
- e) *Nine Windows*
- f) *DTC Operator*
- g) *Functional Modeling and Trimming*
- h) *Subversion Analysis*
- i) *Anti-system*

Solutions Tools
  a) *Four Classical TRIZ Tools*
      o *Scientific Effects (1)*
      o *Inventive Standards (2)*
      o *Contradictions*
          ▪ *Technical Contradiction (3)*
          ▪ *Physical Contradiction (4)*
  b) *Tools of ARIZ*
      o *Tool 1 Form Su-field models use inventive standards*
      o *Tool 2 Separate useful and harmful actions in time and space*
      o *Tool 3 Smart Little People*
      o *Tool 4 Substance-Field Resource Inventory and IFR-1*
      o *Tool 5 Substance-Field Resource Inventory and IFR-2 (physical contradiction at macro and micro levels).*
  c) *Trends of Evolution*

5.2 Evaluation of TRIZ Tools that may be adapted to be useful for finding root cause

So which tools can be used to help find root cause?

Analytical tools

- **S-curve analysis** is used to predict and manage development of a technical system along the S-curve. It is not useful for root cause analysis. ☒

- **Trends of Evolution** predict future technology developments, not useful for root cause analysis. ☒

- **CEC analysis** is a root cause analysis tool. ☑

- **The Ideal solution/system**. The ideal solution to our bleeding finger is there is no bleeding; it does not help with root cause analysis. ☒

- **Nine Windows** can be used for root cause analysis. Was the problem caused at system level, supersystem, in the past or future? ☑

- **DTC Operator.** Changing dimensions, time, and cost may help with root cause. What if we exaggerate not only the dimensions, time and cost to very large and very small but apply this to our Substance-Field Resource list. For example, if the temperature were very high, how could this cause the problem? ☑

- **Functional Modeling and Trimming**. Not trimming but building a functional model can reveal interactions and substances that we could otherwise overlook and is useful for root cause analysis. ☑

- **Subversion Analysis (SA-1).** By trying to think of how to cause the problem may help reveal the actual cause. ☑
- **Anti-system**. Not useful for helping find root cause. ☒

Solutions Tools
- **Scientific Effect**. Researching how a **harmful function or action** can be performed can help lead us to what the cause may be. The problem is expressed as a harmful **function statement.** ☑
- **Inventive Standards**. Determining how a harmful interaction occurs may reveal root cause by stating the Su-field interaction as a function statement and searching scientific effects (above). Inventive standards themselves don't support ideas for root cause analysis. ☒
- **Contradictions.** Contradictions help us recognize what is stopping a solution from being implemented, such that we define a solution that overcomes the opposition and allows the solution to be implemented. We first must know the problem. ☒
- **Tools of ARIZ**
  (Tool 1) **Su-field Modeling and Inventive Standards**: useful only in terms of how to produce the harmful function (mentioned above). ☒
  (Tool 2) **Solutions in Time and Space**: is the problem caused at a different time or space than we think? This is similar to Nine Windows and its use as a root cause tool is captured by Nine Windows. So we will not use it in our list below, we will use Nine Windows instead. ☒
  (Tool 3) **Smart Little People**: Although we draw the Smart Little People performing useful and harmful actions at stages as we transition through a contradiction from one state to the other, we can use the idea of Smart Little People to prompt ideas for how to cause the problem to try to identify root cause. ☑
  (Tool 4) **Substance-Field Resource** list and state the technical contradiction in the form of the **Ideal Final Result** 1 (IFR-1). Although created for solving a technical contradiction, we can build the Substance-Field Resource list of "raw" materials and include combinations, derivatives, emptiness and "hidden resources" and convert the IFR-1 statement to a RCS (root cause statement), an "inverse" IFR-1 statement. ☑
    - **RCS**: <X-resource> causes <the harmful affect>. ☑
  (Tool 5) Resource Inventory and Conversion to a **Physical Contradiction** (IFR-2) Not useful for root cause analysis. ☒

CEC Analysis is already used. So we conclude that the most useful tools we can adapt and add for root cause determination are:

- Nine Windows
- DTC Operator

268

- Functional Modeling
- SA-1 Subversion Analysis 1
- Function Statement and Scientific Effects
- Smart Little People
- SFR list and Root Cause Statement RCS (inverse IFR-1)

Although they may be used in any order, we recommend the following sequence:
- Function Statement and Scientific Effects
- Nine Windows
- Functional Modeling
- SFR list and Root Cause Statement RCS (inverse IFR-1)
- DTC Operator
- Smart Little People
- SA-1 Subversion Analysis 1

Let's discuss how to use them for root cause analysis.

6. TRIZ Tools that can be leveraged to determine root cause and how to apply them to find root cause

Before applying TRIZ tools to determine root cause, the standard steps should be completed. Draw the problem, Perform CEC-1 (and brainstorm and Fishbone if needed). See SSPS Incorporating TRIZ tools for Root Cause Analysis Flowchart in Section 7.1 below.

6.1 Function statement/Scientific effects

When root cause is unknown, all we know is that there is a bad effect. "Something" causes a harmful action or effect to occur. We can state this as a general Subject-Action-Object function statement (or Su-field interaction).

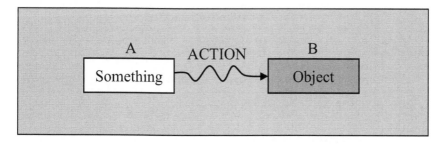

We will use the term "harmful" action but it may also be insufficient, excessive or absent.

The "something" may an object, field or a mechanism made up of many interactions. The action may be permanent fail or may be intermittent, occurring only in specific circumstances. The harmful action may act on a single or multiple objects, but we can summarize the entire failure, the unwanted effect in terms of a single A acts on B (harmful) function statement. For example:

Something corrodes metal
Something forms residue
Something ignites vapor

We can then formulate the problem as: how to perform the harmful action?

By making an inventory of all the resources we can create models of the root cause. We can search for **scientific effects** that could cause the harmful function using the available resources (including hidden, derived etc).

Algorithm for using a Function Statement to find root cause

Step 1: Define the problem as a harmful Subject Action Object action (e.g. something corrodes metal).
Step 2: Search scientific effects for the general function (e.g. how to corrode, corrosion, disintegrate, dissolve, etc.).
Step 3: List all resources. (e.g. metal, air, air contaminants (water vapor, salt, temperature, static charge, etc.)).
Step 4: Create models of how the specific problem can be caused using the resources (e.g. the metal is corroded by the presence of moisture and chlorine in the local atmosphere).

6.2 Nine Windows

Consider different levels of Nine Windows. Sometimes the cause is quickly recognized when we do not focus on the present system level. Something in the supersystem may be the cause (for example ambient temperature was reduced and the occurrences of corrosion increases).

Algorithm for using Nine Windows to find root cause

1. Draw Nine Windows.
2. Write the problem in the center box and complete the diagram by entering the subsystem, system and supersystem components.
3. Consider various timescales and levels of system consider whether the problem is caused at that level/timeframe. Although a problem may be caused in the present or past, sometimes it occurs after the time we think it did.

6.3 Functional Model

Build a model of how the system operates. If it is a complex system with many components, try to focus on the area where the problem is believed to be occurring. Building a functional model of the system is similar to drawing the problem (it is a functional rather than a physical drawing). It can release psychological inertia. We consider many interactions and components we might otherwise miss; we remember to consider interactions with the supersystem. Often we see the problem in a new way that can lead to an idea about root cause.

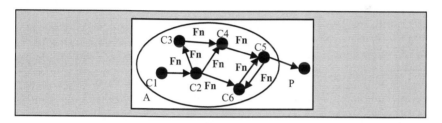

A functional model of components or a process model of steps can be created.

Algorithm for using Functional Modeling to find root cause

Step 1 Determine what is to be modeled, the system or the specific part of the system where the problem resides.
Step 2 Build the model of how the system (or chosen parts) or process (or chosen steps) should operate.
Step 3 Create ideas for how the problem can be caused.

6.4a Substance-Field Resources (SFR list)

The Substance-Field Resource list should contain all the substances, fields and their parameters. Unlike a standard SFR list, we don't define a tool or affected object (we are not analyzing a "system" that has a technical contradiction). We simply make an inventory of all of the components including the supersystem. Note that it is necessary to add the derivatives; hidden components (also list any gaps, spaces that occur). This is because if the problem has not been identified using raw resources, the cause is often a hidden, derived or a combination of resources which are often overlooked.

When creating the SFR list, consider the analytical and experimental data and frequency of occurrence. If the problem turned on at a specific time, consider resources that may have changed in the estimated timeframe the problem started. This will minimize the potential resources to consider.

Substance-Field Resource Analysis

1. System Resources (Internal) – resources of the system
  - Components

2. Supersystem Resources (External)
  - SFR resources of the operating space environment at the problem
  - SFR resources of common to any external environment (e.g. gravity, the earth's magnetic field, air etc).

3. Surrounding Resources
  - Waste material from outside the system, by products, nearby systems and resources
    Note that we exclude cheap or inexpensive components that could be used to fix a problem; we are searching for root cause.

Raw Resources

SFR List			
Internal System Resources	Substance	Field	Parameter
1 System Resources			
System components			
External Resources			
2 Supersystem Resources			
Environment where the problem is believed to reside			
Common to any environment			
3 Surrounding Resources			
By products, waste material			
Nearby systems and resources			

If the above "raw" resources do not provide a solution, try the following:

Secondary Resources
Use a mixture of resources.

Use a mixture of resources	

Consider using empty space or a combination of a substance with empty space to solve the problem (empty space can mean space, gap, cavities, bubbles etc).

Use empty space (gaps, bubbles, cavities, voids etc.).	

Consider using derivatives (or a mixture of derivatives with empty space)

A derivative is a changed resource, changed **phase**, part of materials, burnt remains, decomposed materials, molecules etc anything we can derive from an existing resource or resources.

Use derived resources	
Use derived resources with empty space **(gaps, bubbles, cavities, voids etc.).**	

Hidden resources These are resources that are present but often overlooked. For example contamination level in water, foreign materials or constituents of a gas (instead of air, nitrogen, oxygen, argon etc, instead of wood, moisture etc).

Use hidden resources	
Use hidden resources **(contaminants, foreign materials, constituents etc.).**	

6.4 b Build the Root Cause Statement (RCS) (this is a similar way we create an IFR-1 statement in ARIZ).

RCS: <X-resource> causes <the harmful affect>.

For example, we can use the example of a plasma etch tool used in the semiconductor industry to etch aluminum interconnecting metal. It comprises an electrode, chamber walls, process, gases, a robot transfer arm etc. An RCS list would look like this:

- The **electrode** causes the corrosion of the metal
- The **chamber walls** cause the corrosion of the metal
- The **metal** causes the corrosion of the metal
- The **temperature of the electrode** causes the corrosion of the metal
- The **pressure of the transfer chamber** causes the corrosion of the metal
- The **process time** causes the corrosion of the metal
- The **motion of the arm** causes the corrosion of the metal
- The **pressure calibration** causes the corrosion of the metal.
- The **environment** causes the corrosion of the metal
- The **ambient light** causes the corrosion of the metal
- The **moisture** causes the corrosion of the metal
- The **contamination level** of the air causes the corrosion of the metal
- The **purity of the process gases** causes the corrosion of the metal

By considering each in turn, we release psychological inertia leading to many new ideas and potential solutions. This is a typical TRIZ "solutions first" approach, instead of searching for a cause, we list all possible solutions first then eliminate the ones that are not the cause.

Theoretically, the RCS statement should contain the root cause (or part of it), given all the available fields, and all the substances, raw, hidden, derived, etc, that are present or can be derived or combined. Only the mechanism (model scenario) is missing.

Algorithm for using Substance-Field Resources (SFR) and Root Cause Statement (RCS) to find root cause

Step 1: Complete the SFR (substance and field resource table, include raw, derived, hidden resource, etc. Note there is no tool or product/objects).
Step 2: Create the Root Cause Statement RCS: <X-resource> causes <the harmful affect>.
Step 3: Substitute each raw resource then each secondary resource into the X-resource part of the RCS statement.
Step 4: Create ideas based on the psychological inertia released by the RCS statement.

6.5 DTC Operator

Changing the DTC operator may help with root cause. What if we exaggerate not only the dimensions, time and cost (DTC) to very large and very small but apply this exaggeration to every substance, field and parameter on our SFR list. For example if we had a complete SFR list for the plasma etch tool and our problem is the corrosion of the metal. We can ask for every item on the raw and secondary SFR list, every substance, field and parameter (property), could it cause the problem if it were at the extremes.

Would the corrosion occur:

- If the electrode was very large? If the electrode was very small? If the electrode were very hot? If the electrode were very cold? If the thickness of the electrode were low? If the thickness of the electrode very high? If the roughness of the electrode were low? If the roughness of the electrode were high? If the chamber walls were very hot? etc.

The responses may lead to root cause identification. For example, low electrode temperature is a known cause of metal (aluminum) corrosion in the manufacture of semiconductor devices.

Algorithm for using Substance-Field Resources (SFR) and DTC Operator to find root cause

274

Step 1: Create an SFR list
Step 2: Apply the extreme "what if" question questions to each SFR raw and secondary resource. Does this suggest a root cause?

6.6 Smart Little People

Although we draw the Smart Little People performing useful and harmful actions at stages through a contradiction as we transition from one state to another, we can use the idea of Smart Little People to prompt ideas for how to cause the problem to try to identify root cause. How would Smart Little People cause the problem? Make a list.

Algorithm for using Smart Little People to find root cause
Step 1: Draw the problem.
Step 2: Create ideas of what Smart Little People (there may be several groups) are doing to cause the problem.
Step 3: Create plausible models of root cause based on what the Smart Little People are doing.

6.7 Subversion Analysis SA-1

Develop (brainstorm) model scenarios for how to *cause* the problem (we do not use CEC analysis; this should have been done in the Standard Structured Problem Solving step). Play devil's advocate. Given all the resources, including the secondary hidden and derived resources, their storage, manufacture, the environment, people, procedures etc, build model scenarios of how can you cause or exacerbate the problem? Initial ideas should be brainstormed, with no need to be consistent with the data or experimental results.

For example: how to cause the metal to corrode?
Ideas:
- Put a leak in the vent pipe, add moisture to the air, decrease the temperature of the electrode, increase the pressure of helium heat transfer medium, etc.

Next evaluate whether any of the models are consistent with the data and experiment results. Then try to determine whether they are the root cause.

Algorithm for using Subversion Analysis SA-1 to find root cause

Step 1: State the problem.
Step 2: List all the resources, including the secondary hidden and derived resources (see above SFR list) including their storage, manufacture, operation, the environment, people, procedures, etc.
Step 3: Given all the resources, create models of how to cause the problem.

275

Step 4: Evaluate whether any of the models are consistent with the data and experiment results. Then try to determine whether they are the root cause.

The above TRIZ tools should help substantially with root cause analysis. After root cause is defined, solutions may be identified without the need for divergent thinking tools. If creative thinking is needed, then apply TRIZ.

7. Algorithm for Standard Structured Problem Solving Incorporating TRIZ tools for Root Cause Analysis

See flowchart and template for Standard Structured Problem Solving Incorporating TRIZ Tools for Root Cause Analysis in Appendices 8.5 and 8.6 for applying the algorithm.

Step 1: Define Problem

Step 2: Current Situation

- **Draw problem** It is useful to try to draw the system where the problem resides, as close to the location as is known. Draw at least both side and top down views. Drawing can often release psychological inertia.

Step 3 Analysis

- Data analysis and results

Step 3.1 Root Cause Unknown

- Brainstorming
- Create model scenarios
- CEC analysis

Step 3.1A If root cause still unknown

- Check data (define plan to validate data is 100% correct)
- Eliminate groupthink

Step 3.1B If root cause STILL unknown

Apply TRIZ Tools adapted for Root Cause Analysis

  - Create Function Statement for harmful effect/Search scientific effects
  - Nine Windows
  - Build Functional Model
  - Create SFR List
    - a) Raw
    - b) Mixture
    - c) Emptiness
    - d) Derived
    - e) Hidden
  - RCS Statements
  - DTC Operator
  - SLP
  - Subversion Analysis SA1

Step 4 Solutions

Step 5 Implementation

276

7.1 SSPS Incorporating TRIZ tools for Root Cause Analysis Flowchart

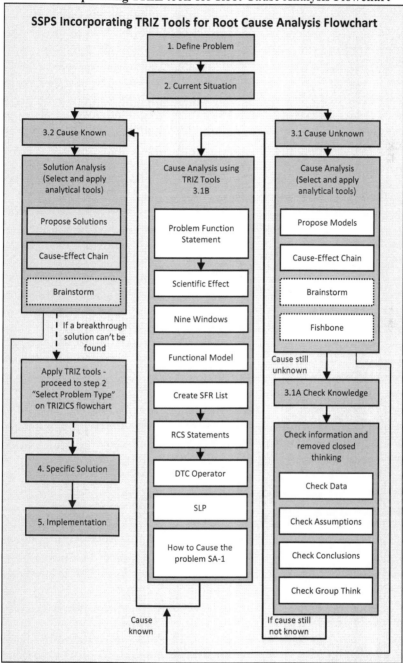

SSPS Incorporating TRIZ Tools for Root Cause Analysis Flowchart

Note that it is not necessary to use all of the analytical tools and the user may choose in which sequence to use them.

277

Glossary

A

Administrative Contradiction: a simple conflict statement that can be expressed in the general form: I have a goal but don't know how to achieve it. For example - I want to do something but I don't know how to do it. The statement does not provide a direction for solving the problem.

Advanced Contradiction Tools: advanced tools for solving a contradiction are the main tools of ARIZ. See basic tools for a description of basic tools for solving a technical and physical contradiction.

- For solving a technical contradiction advanced tools are: Tool 1: Su-field Modeling and inventive standards, Tool 2: Solutions in Time and Space (technical contradiction), Tool 3: Use Smart Little People, Tool 4: State the technical contradiction in the form of the Ideal Final Result 1 (IFR-1).
- Advanced tools for solving a physical contradiction are formulating the Ideal Final Result-2 IFR-2 and using satisfaction separation and bypass.

Altshuller: Genrikh Saulovich Altshuller, 1926-1998 b. Uzbekistan, USSR. The creator of TRIZ and science fiction writer - penname Genrikh Altov.

Altshuller's Matrix (see contradiction the Contradiction Matrix)

Analogous thinking: the ability to create specific solutions that are conceptually alike to a general solution. For example, to create specific ideas prompted by an inventive principle or inventive standard.

Analytical tool: a thinking tool used to clarify or simplify a problem so it is easier to solve. After applying analytical tools the problem is redefined as one or more specific problems to solve.

Anti-system: a tool for releasing psychological inertia by considering the anti-function or anti-action or a system. It is sometimes used with Nine Windows.

ARIZ: is the acronym for algorithm for inventive problem solving (Algorithm Rezhenija Izobretatelskih Zadach) also known as ASIP (Algorithm for Solving Inventive problems) in English. A multi-step sequential algorithm for inventive problem solving, it starts by stating the problem as a technical contradiction (hence it defines an "inventive problem") and applies a number of solutions tools to it. The process later states the problem as a physical contradiction and applies a number of additional tools to create solutions. The process is recommended for difficult problems. Altshuller estimated around 15% of inventive problems may require ARIZ. Using ARIZ exposes the problem to many solutions tools. There are many versions of ARIZ; the last "official" version approved by Altshuller is ARIZ-85C, see Appendix 8.9. See also Simplified ARIZ Chapter 9.

B

Basic Contradiction Tools: the basic methods for solving a Technical Contradiction are Altshuller's Matrix and the 40 Principles. The basic method for solving a physical contradiction is to identify antonym pairs that support opposing requirements and use satisfaction, separation and bypass. Most inventive problems can be solved using basic tools. If there is no solution found or a more innovative solution is required then we recommend that user applies the advanced tools for solving a contradiction by following the Contradiction Problem Solving Flowchart – Simplified ARIZ.

Brainstorm: a method of creative thinking used to find root cause or identify solutions. Typically performed as a group activity, it is a process for gathering creative ideas to solve a problem. The target problem is stated, potential causes or solutions are identified by moving sequentially around the room to get ideas from individuals. All are encouraged to use their creative mind to generate ideas. Of key importance is that ideas are generated with no criticism to try to free the less logically derived ideas and to open the imagination. The idea is that the most radical solutions are derived towards the end of such a session. When few ideas are being created the more outrageous come to the fore and these are what we are trying to find. Brainstorming has a more colloquial meaning which is to create a list of ideas.

Bypass: a method of solving a physical contradiction problem by moving to a completely different method that avoids the physical contradiction.

C

CEC: Cause-Effect Chain analysis: a form of logically directed brainstorming. It can be used to find root cause, to identify the correct problem is being addressed, to create general ideas for improvement and to identify possible failure modes.

Classical TRIZ tools: classical TRIZ comprises four main "solutions tools": scientific effects (a list of scientific phenomena that solve a problem, the problem is formulated as a search of a function), inventive standards (a list of 76 ways of improving the functionality of an interaction, the problem is formulated as a Su-field model), technical contradiction (a dilemma usually formed via an "if-then-but" statement solved by inventive principles or the advanced tools of ARIZ including inventive standards) and a physical contradiction (formulated by forming antonym pairs or the advanced tools of ARIZ and solved by satisfaction, separation or bypass). Sometimes the laws or trends of evolution are used as a classical TRIZ solutions tool by reviewing trends to prompt ideas.

Component: part of a system that provides a function. In functional modeling this can be a physical object or a field.

Containment Plan: a short term plan to mitigate loss or damage while a permanent solution is sought. When a reactive problem occurs, the first step is often to introduce a containment plan.

Contradiction: TRIZ defines three forms see, administrative, technical and physical contradictions.

Contradiction Diagram: a powerful tool for analyzing a technical contradiction. From the "if-then-but" statement the diagram is used to define: the states of the two technical conflicts TC-1 and TC-2, the improving and worsening parameters, the parameter of change, the tool and the useful and harmful actions in each state. With this information the problem can be easily reformulated, allowing application of the basic and advanced tools for solving a contradiction.

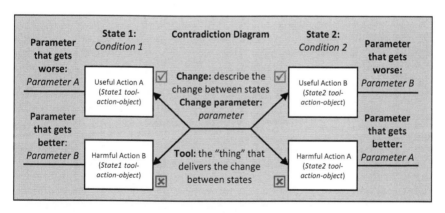

Contradiction Matrix (also known as Altshuller's Matrix): a basic tool for solving technical contradictions. A 39 by 39 cell table used to solve a technical contradiction, improving and worsening general parameters are identified by the user, the table is used to reference up to four of the most commonly used of 40 inventive principles to solve the conflict.

D

Derived Resources: a derivative is a changed resource, changed phase, parts of materials, burnt remains, decomposed materials, molecules, etc. Anything we can derive from an existing resource or resources.

Drawing: a powerful tool for releasing Psychological Inertia is to draw the problem. Often items that can cause or solve the problem are overlooked and are revealed by a drawing. It is useful to draw top, front, side and angled views.

DTC Operator: Dimension Time Cost Operator. A tool for releasing psychological inertia by imagining if size, time (including speed), or cost were exaggerated to extremes.

F

Fishbone Diagram: ideas are organized by similar categories using a diagram that resembles a fishbone. Used to find the root cause of a problem - the source may be easier to find when broken down into specific categories.

Functional Model (of a technical system or technical process): a functional model illustrates how a system functions, the functions performed by components, and how they interact with the product and supersystem are powerful for releasing psychological inertia. In a similar way to drawing, a functional model often reveals the cause or solution to a problem that is overlooked.

G

Graphical Representation: a diagram of the tool and object(s) showing the useful and harmful actions in technical conflict 1 and technical conflict 2 for a technical contradiction.

H

Harmful Action or Function: an undesired action or function that produces a negative impact. Often absent, insufficient or excessive actions or functions are referred to as the harmful function for convenience when compared to the useful function. In functional modeling harmful, absent, insufficient and excessive are defined specifically and not all referred to a harmful.

Hidden resources: resources that are present but often overlooked due to Psychological Inertia. For example, the contaminant level in water or air, the static charge on a sheet of material.

282

Ideality: technical systems drive towards becoming more ideal, they evolve to increase functionality, decrease in cost and harmful effects.

Ideal Solution: a solution that solves itself, for free. Used to release psychological inertia, how can the problem solve itself? This is similar to, but should not be confused with Ideal Final Result IFR-1 and IFR-2.

Ideal System: a system that has infinite functionality at zero cost or harm. The ideal process is one that produces the output for free. The ideal maintenance procedure is no maintenance procedure.

IFR-1: an ARIZ term for Ideal Final Result 1. It is a way of stating a technical contradiction in terms of opposing actions. A statement is created for each resource, combinations of resources, etc. IFR-1 is actually a list of statements rather than one. It takes the form: "X-resource **itself** provides <the useful action> and eliminates <the harmful action> without making the system more complex and without additional harmful consequences."

IFR-2: an ARIZ term for Ideal Final Result 2. It is a way of stating the physical contradiction. Like IFR-1, the IFR-2 statement is created for each resource, combinations of resources, etc. and is a list of statements rather than one. Multiple IFR-2's can be created for a single resource. The IFR-2 is a combined list of macro physical contradiction statements and micro level (the level of the very small, particles, atoms, etc.) physical contradiction statements.

"If-then-but" statement: basic formulation of a technical contradiction. It takes the general form: if (state what change is being made), then (state what improves) but (state what gets worse). Typically, this is the first step used to form a contradiction is to form the "if-then-but" statement.

Inventive Principle: a list of 40 conceptual ideas that have been used to solve technical contradictions.

Inventive Problem: a problem that involves solving a contradiction, and therefore requires breakthrough thinking, classed as problem at Level 2 and up in terms of the Level of Invention.

Inventive Standard: a list of 76 "standard" solutions for solving a problem when it is expressed as an interaction between two substances (physical objects) and a field. Used also as solutions to technical contradictions by addressing the harmful (negative) action of a technical conflict (TC-1 or TC-2).

L

Level of Invention: Altshuller classified patents into 5 Levels of Inventiveness increasing in difficulty from 1 to 5. 1 being a basic routine improvement and 5 which involves highly innovative thinking that requires new scientific discovery.

M

Main Useful Function (MUF): the main function performed by a technical system or process. The main purpose of why the system was created.

MATCHEM: a mnemonic used to identify basic types of field that increase in terms of evolutionary development from Mechanical, to Acoustic, Thermal, Chemical, Electric then to Magnetic (or Electromagnetic). Mechanical is considered least evolved and least controllable, progressing through the list to magnetic or electromagnetic which is considered the most evolved and most controllable. The trends of evolution states basic fields develop towards more evolved fields.

Matrix Parameter: a list of 39 typical features or characteristics of a system that improves or worsens as a change is made to the opposite state of a technical contradiction.

N

Nine Windows (also known as Nine Screens or System Operator): a creative thinking tool that ensures we consider our search for the cause or solution to a problem not only at the present system level but also in the past and future and at subsystem and supersystem levels.

Null action: an action in a technical contradiction that is considered good because it is not performed. For example, chemical does not dissolve metal is a null action. On the opposite state of the contradiction the negative action is chemical dissolves metal (harmful).

O

Operating Space: The area of space in which a conflict occurs. It may change depending on the operating time T1 to T2 to T3.

Operating Time: The time of conflict (TOC), the time when a conflict occurs. We break the contradiction into three separate problem situations to help the problem solving process, T1, T2 and T3. The TOC is T1, T2 and T3.

- T- the time in advance of State 1

284

- T1 the time during TC-1 (State 1)
- T2 the time during transition between states.
- T3 the time during TC-2 (State 2)
- T+ the time after State 2

P

Parameter: a characteristic, feature or property of a system. For example, shape, temperature, flexibility, etc. A matrix parameter is one of 39 general parameters. A specific parameter is categorized as one of the general parameters in order to use the Contradiction Matrix.

Physical Contradiction: a physical contradiction is formed when a system, component, part of a component, etc. has opposing physical requirements. The system, component, part of a component must have two opposite states, features, parameters, etc. X must be "A" and "anti-A" where "A" and "anti-A" are opposite states features, parameters, properties, etc. For example, the coffee must be hot and cold, the pencil must be sharp and blunt, the electric charge must be present and not present.

Problems Bank: as the problem solving process proceeds, a number of specific problems are identified that, if solved, could solve the main target problem. The specific problems identified are compiled in one list. At the start of the problem solving process, a blank table for capturing ideas of both problems and possible solutions/ideas are created and these are compiled as the process proceeds.

Problem Type: problems are classified into four types - needed for directing the user to the appropriate analytical tools. The problem is fitted to one of these types - specific, cause unknown (1), specific, cause known (2), non-specific, general improvement (3), and failure prevention (4).

Product: The object of the main useful function in a subject-action-object statement

Psychological Inertia (also known as Mental Inertia): the resistance to free creative thinking due to unconsciously imposed barriers.

R

Resource (see also SFR): A physical object, material, field, parameter that is available or can be made available to be used to provide a solution.

Root Cause: the fundamental underlying reason for a problem.

Root Cause Statement (RCS): used to release psychological inertia when trying to find root cause. An inventory of resources is compiled and each resource (including, derived, hidden, combinations, etc.) is input into the RCS statement: <X-resource> causes <the harmful affect>.

S

Satisfaction: the system or component is directly endowed with a property that can satisfy conflicting requirements of a physical contradiction.

Scientific Effect (also known as physical effect): a scientific effect or phenomenon (physical, chemical, biological, geometrical, etc.) that can be used to solve the problem.

S-curve: a curve describing the typical evolution of the main parameters of a system over time.

Secondary Problem: a new problem that is created by implementing the solution to the original problem.

Separation: the conflicting requirements of a physical contradiction can be solved by separating them in time, space, relation or system level.

Simplified ARIZ: a method for solving inventive problems that incorporates the basic tools and the key advanced tools of ARIZ for solving a contradiction. It is performed in four phases A, B, C and D.

A. Form and Solve Technical Contradiction Algorithm using Basic Tools
B. Form and Solve the Physical Contradiction Algorithm using Basic Tools
C. Form and Solve the Technical Contradiction using Advanced Tools
D. Form and Solve the Physical Contradiction using Advanced Tools

Smart Little People (also known as many little men, etc.): imaginary little people used as a thinking tool to release psychological inertia. The little people are endowed with various imaginary powers or capabilities to describe and solve a problem. There can be many different groups of little people in different quantities with different featured and "powers." SLP is most commonly used to model the useful and harmful actions performed of a technical contradiction. The actions performed by the little people are used to prompt ideas by analogous thinking.

Solutions Bank: as the problem solving process proceeds, solutions or ideas can arise at any time. The specific solutions identified are compiled in one list. At the start of the problem solving process, a blank table for capturing ideas of both problems and possible solutions/ideas are created and these are compiled as the process proceeds.

286

Solutions Tool: a tool for creating ideas for solving specific known problems. The solutions tools are the classical TRIZ tools plus reviewing the trends which may prompt an idea.

Standard Structured Problem Solving: standard organized method for solving problems that focuses mainly on "in-the-box" thinking.

Subject-Action-Object Interaction (SAO): a basic form of describing an action or function. The subject acts upon the object such that one or more parameters of the object are changed.

Substance: in Su-field modeling substance means "thing" any physical object or material.

Substance-Field Resource List (SFR List): an inventory of all substances (things), fields and their parameters that are available or may be made available to solve the problem.

Subsystem: a component or sub-part of a system.

Subversion Analysis: there are two types. SA-1 (Subversion Analysis 1): used to reveal the root cause of a (reactive) problem by trying to cause it. SA-2 (Subversion Analysis 2): used to try to identify the ways a system might fail (failure modes) in future by thinking of ways failures may be caused.

Su-field (also known as Substance–Field interaction): a complete Su-field is one in which an object acts upon another via a field. In the same way a subject-action-object interaction takes place.

Supersystem: the external components a system interacts with or the associated components or systems that are externally related to a system. When dealing with a functional model the supersystem is made up of external elements not designed as part of the system but interact or may interact with it. If the system is a car, the supersystem is made up of elements such as gravity, rain, roads, the driver (who is also the product), air, light, the environment. We can also consider broader aspects of the supersystem related to the car, gas stations, bridges, pedestrians, other traffic, are also part of the car supersystem. We can continue expanding the definition of the supersystem further - oil refineries, ships for moving oil, forests for growing rubber, farms to supply leather. The supersystem list can continue to grow. The appropriate expanded limits to use when defining the supersystem elements depend upon the desired solution and are problem specific.

System: a technical system is a system that was created to perform one or more functions.

287

T

Technical Conflict: one half of a technical contradiction (normally denoted as TC-1 or TC-2).

Technical Contradiction: a dilemma formed by an "if-then-but" statement that defines two technical conflicts, TC-1 in State 1 and TC-2 in State 2. A technical contradiction can be expressed in terms of opposing parameters and as opposing actions. We try to make an improvement so that "something good" happens but "something bad" happens as well.

Technical Process: A procedure or process that is made up of steps that perform one or more functions.

Technical System: an object or combination of objects that is created to perform one or more functions.

Time of Conflict (T.o.C): see operating time, Time of conflict is T1, T2 and T3 when the contradiction occurs.

Tool: the object that delivers the change between states in a technical contradiction.

Tools of ARIZ: the ARIZ algorithm disassembled into individual problem solving tools that can be applied without the need to use entire ARIZ sequence of steps.

Trends of Evolution (also known as Laws of Evolution): a recognized repeating pattern of how systems evolve.

Trial and Error: a method of problem solving based on trying numerous solutions with no clear direction to derive the solution, similar to guessing, hit and miss, etc. TRIZ is not trial an error but bases problem solving on data derived from analysis of patents and tools designed for directing and releasing the creative imagination.

Trimming: removing components of a system and allocating their function to another component or part of the system or supersystem.

TRIZ: the acronym of Teoriya Resheniya Izobreatatelskikh Zadatch, also known as TIPS in English, Theory of Inventive Problem Solving.

TRIZICS: the method we use for problem solving using TRIZ. It has six phases 1. Identify Problem, 2. Select Problem Type, 3. Apply Analytical Tools, 4. Define Specific Problem, 5. Apply TRIZ Solutions Tools, 6. Solutions and Implementation.

288

U

Useful action or function: an action performed by one object on another that is considered to be wanted or desired.

V

Variant: an idea for a solution. An empty variant is an idea that is unsuccessful.

Z

Zone of Conflict (Z.o.C) another name for Operating Space.

Appendices

1. *The 39 Parameters and 40 Inventive Principles*
2. *Contradiction Matrix*
3. *The 40 Principles with Examples*
4. *The 39 Parameters Definitions*
5. *Inventive Standards Flowchart*
6. *The 76 Inventive Standards*
7. *Table of Specific Inventive Principles to Solve Physical Contradictions.*
8. *Flowcharts/Roadmaps/Templates*
   8.1. *TRIZICS Roadmap*
   8.2. *TRIZICS Problem Solving Template*
   8.3. *Standard Structured Problem Solving Flowchart*
   8.4. *Standard Structured Problem Solving Template*
   8.5. *Standard Structured Problem Solving Incorporating TRIZ tools for Root Cause Analysis Flowchart*
   8.6. *Standard Structured Problem Solving Incorporating TRIZ Tools for Root Cause Analysis Template*
   8.7. *Contradiction Problem Solving Flowchart – Simplified ARIZ*
   8.8. *Simplified ARIZ Algorithm*
   8.9. *ARIZ-85C*

Appendix 1. The 39 Parameters and 40 Inventive Principles

A list of the 39 typical parameters that can be in technical conflict (improving one causes the other to worsen) and the 40 ideas (principles) used to solve a technical conflict.

	39 Parameters		40 Principles
1	Weight of moving object	1	Segmentation
2	Weight of stationary object	2	Taking out / Extraction
3	Length of moving object	3	Local Quality
4	Length of stationary object	4	Asymmetry
5	Area of moving object	5	Merging / Combination
6	Area of stationary object	6	Universality
7	Volume of moving object	7	Nested Doll
8	Volume of stationary object	8	Anti-weight / Counter-weight
9	Speed	9	Prior counter-action
10	Force (Intensity)	10	Preliminary action / Prior action
11	Stress or pressure	11	Beforehand cushioning
12	Shape	12	Equi-potentiality / Remove tension
13	Stability of composition	13	The other way round
14	Strength	14	Spheroidality-Curvature
15	Duration of action of moving object	15	Dynamics
16	Duration of action by stationary object	16	Partial or excessive actions
17	Temperature	17	Another Dimension
18	Illumination intensity	18	Mechanical Vibration
19	Use of energy by moving object	19	Periodic action
20	Use of energy by stationary object	20	Continuity of useful action
21	Power	21	Skipping / Hurrying
22	Loss of energy	22	Blessing in Disguise
23	Loss of substance	23	Feedback
24	Loss of Information	24	Intermediary
25	Loss of time	25	Self- Service
26	Quantity of substance	26	Copying
27	Reliability	27	Cheap / Short Living
28	Measurement accuracy	28	Mechanics substitution / Another sense
29	Manufacturing precision	29	Pneumatics and hydraulics / Fluidity
30	Object-affected harmful factors	30	Flexible shells and thin films
31	Object-generated harmful factors	31	Porous Materials / Holes
32	Ease of manufacture	32	Color changes
33	Ease of operation	33	Uniformity
34	Ease of repair	34	Discarding and recovering
35	Adaptability or versatility	35	Parameter changes
36	Device complexity	36	Phase transitions
37	Complexity of control	37	Thermal expansion / Relative change
38	Extent of automation	38	Strong oxidants / Enriched atmosphere
39	Productivity	39	Inert atmosphere / Calmed atmosphere
	----------	40	Composite Materials/ Structures

Appendix 2. Contradiction Matrix

291

Improving Parameter ↓ / Worsening Parameter →	Weight of moving object	Weight of stationary object	Length of moving object	Length of stationary object	Area of moving object	Area of stationary object
	1	2	3	4	5	6
1 Weight of moving object	+	-	15,8, 29,34	-	29,17, 38,34	-
2 Weight of stationary object	-	+	-	10,1, 29,35	-	35,30, 13,2
3 Length of moving object	8,15, 29,34	-	+	-	15,17, 4	-
4 Length of stationary object		35,28, 40,29	-	+	-	17,7, 10,40
5 Area of moving object	2,17, 29,4	-	14,15, 18,4	-	+	-
6 Area of stationary object	-	30,2, 14,18	-	26,7, 9,39	-	+
7 Volume of moving object	2,26, 29,40	-	1,7, 4,35	-	1,7, 4,17	-
8 Volume of stationary object	-	35,10, 19,14	19,14	35,8, 2,14		-
9 Speed	2,28, 13,38	-	13,14, 8	-	29,30, 34	-
10 Force (Intensity)	8,1, 37,18	18,13, 1,28	17,19, 9,36	28,10	19,10, 15	1,18, 36,37
11 Stress or pressure	10,36, 37,40	13,29, 10,18	35,10, 36	35,1, 14,16	10,15, 36,28	10,15, 36,37
12 Shape	8,10, 29,40	15,10, 26,3	29,34, 5,4	13,14, 10,7	5,34, 4,10	-
13 Stability of composition	21,35, 2,39	26,39, 1,40	13,15, 1,28	37	2,11, 13	39
14 Strength	1,8, 40,15	40,26, 27,1	1,15, 8,35	15,14, 28,26	3,34, 40,29	9,40, 28
15 Duration of action of moving object	19,5, 34,31		2,19, 9	-	3,17, 19	-
16 Duration of action by stationary object	-	6,27, 19,16	-	1,40, 35	-	-
17 Temperature	36,22, 6,38	22,35, 32	15,19, 9	15,19, 9	3,35, 39,18	35,38
18 Illumination intensity	19,1, 32	2,35, 32	19,32, 16	-	19,32, 26	-
19 Use of energy by moving object	12,18, 28,31	-	12,28	-	15,19, 25	-
20 Use of energy by stationary object	-	19,9, 6,27	-	-	-	-
21 Power	8,36, 38,31	19,26, 17,27	1,10, 35,37		19,38	17,32, 13,38
22 Loss of energy	15,6, 19,28	19,6, 18,9	7,2, 6,13	6,38, 7	15,26, 17,30	17,7, 30,18
23 Loss of substance	35,6, 23,40	35,6, 22,32	14,29, 10,39	10,28 24	35,2, 10,31	10,18, 39,31
24 Loss of Information	10,24, 35	10,35, 5	1,26	26	30,26	30,16
25 Loss of time	10,20, 37,35	10,20, 26,5	15,2, 29	30,24, 14,5	26,4, 5,16	10,35, 17,4
26 Quantity of substance/the matter	35,6, 18,31	27,26, 18,35	29,14, 35,18	-	15,14, 29	2,18, 40,4
27 Reliability	3,8, 10,40	3,10, 8,28	15,9, 14,4	15,29, 28,11	17,10, 14,16	32,35, 40,4
28 Measurement accuracy	32,35, 26,28	28,35, 25,26	28,26, 5,16	32,28, 3,16	26,28, 32,3	26,28, 32,3
29 Manufacturing precision	28,32, 13,18	28,35, 27,9	10,28, 29,37	2,32, 10	28,33, 29,32	2,29, 18,36
30 Object-affected harmful factors	22,21, 27,39	2,22, 13,24	17,1, 39,4	1,18	22,1, 33,28	27,2, 39,35
31 Object-generated harmful factors	19,22, 15,39	35,22, 1,39	17,15, 16,22	-	17,2, 18,39	22,1, 40
32 Ease of manufacture	28,29, 15,16	1,27, 36,13	1,29, 13,17	15,17, 27	13,1, 26,12	16,40
33 Ease of operation	25,2, 13,15	6,13, 1,25	1,17, 13,12	-	1,17, 13,16	18,16, 15,39
34 Ease of repair	2,27, 35,11	2,27, 35,11	1,28, 10,25	3,18, 31	15,13, 32	16,25
35 Adaptability or versatility	1,6, 15,8	19,15, 29,16	35,1, 29,2	1,35, 16	35,30, 29,7	15,16
36 Device complexity	26,30, 34,36	2,26, 35,39	1,19, 26,24	26	14,1, 13,16	6,36
37 Complexity of Control	27,26, 28,13	6,13, 28,1	16,17, 26,24	26	2,13, 18,17	2,39, 30,16
38 Extent of automation	28,26, 18,35	28,26, 35,10	14,13, 17,28	23	17,14, 13	-
39 Productivity	35,26, 24,37	28,27, 15,3	18,4, 28,38	30,7, 14,26	10,26, 34,31	10,35, 17,7

	Volume of moving object	Volume of stationary object	Speed	Force (Intensity)	Stress or pressure	Shape	Stability of composition	Strength	Duration of action of moving object
	7	**8**	**9**	**10**	**11**	**12**	**13**	**14**	**15**
1	29, 2, 40, 28	-	2, 8, 15, 38	8, 10, 18, 37	10,36, 37, 40	10, 14, 35, 40	1, 35, 19, 39	28, 27, 18, 40	5, 34, 31, 35
2	-	5, 35, 14, 2	-	8, 10, 19, 35	13, 9, 10, 18	13,10, 29, 14	26,39, 1, 40	28, 2, 10, 27	-
3	7, 17, 4, 35	-	13, 4, 8	17, 10, 4	1, 8, 35	1, 8, 10, 29	1, 8, 15, 34	8, 35, 29, 34	19
4	-	35, 8, 2,14	-	28, 10	1, 14, 35	13,14, 15, 7	39, 37, 35	15, 14, 28, 26	-
5	7, 14, 17, 4	-	29, 30, 4, 34	19, 30, 35, 2	10, 15, 36, 28	5, 34, 29, 4	11, 2, 13, 39	3, 15, 40, 14	6, 3
6	-	-	-	1, 18, 35, 36	10,15, 36, 37	-	2, 38	40	-
7	+	-	29, 4, 38, 34	15,35, 36, 37	6, 35, 36, 37	1, 15, 29, 4	28,10, 1, 39	9, 14, 15, 7	6, 35, 4
8	-	+	-	2, 18, 37	24, 35	7, 2, 35	34,28, 35, 40	9, 14, 17, 15	-
9	7, 29, 34	-	+	13,28, 15, 19	6, 18, 38, 40	35,15, 18, 34	28,33, 1, 18	8, 3, 26, 14	3, 19, 35, 5
10	15, 9, 12, 37	2, 36, 18, 37	13,28, 15, 12	+	18, 21, 11	10,35, 40, 34	35, 10, 21	35, 10, 14, 27	19, 2
11	6, 35, 10	35, 24	6, 35, 36	36, 35, 21	+	35, 4, 15, 10	35,33, 2, 40	9, 18, 3, 40	19, 3, 27
12	14, 4, 15, 22	7, 2, 35	35, 15, 34, 18	35, 10, 37, 40	34, 15, 10, 14	+	33, 1, 18, 4	30, 14, 10, 40	14, 26, 9, 25
13	28, 10, 19, 39	34,28, 35, 40	33,15, 28, 18	10,35, 21, 16	2, 35, 40	22, 1, 18, 4	+	17, 9, 15	13, 27, 10, 35
14	10,15, 14, 7	9, 14, 17, 15	8, 13, 26, 14	10,18, 3, 14	10, 3, 18, 40	10,30, 35, 40	13, 17, 35	+	27, 3, 26
15	10, 2 19, 30	-	3, 35, 5	19, 2, 16	19, 3, 27	14,26, 28, 25	13, 3, 35	27, 3, 10	+
16	-	35, 34, 38	-	-	-	-	39, 3, 35, 23	-	-
17	34,39, 40, 18	35, 6, 4	2, 28, 36, 30	35,10, 3, 21	35,39, 19, 2	14,22, 19, 32	1, 35, 32	10, 30, 22, 40	19, 13, 39
18	2, 13, 10	-	10, 13, 19	26, 19, 6		32, 30	32, 3, 27	35, 19	2, 19, 6
19	35, 13, 18	-	8, 35, 35	16,26, 21, 2	23, 14, 25	12, 2, 29	19,13, 17, 24	5, 19, 9, 35	28, 35, 6, 18
20	-	-	-	36, 37	-	-	27, 4, 29, 18	35	-
21	35, 6, 38	30, 6, 25	15, 35, 2	26, 2, 36, 35	22, 10, 35	29, 14, 2, 40	35, 32, 15, 31	26, 10, 28	19, 35, 10, 38
22	7, 18, 23	7	16, 35, 38	36, 38	-	-	14, 2, 39, 6	26	-
23	1, 29, 30, 36	3, 39, 18, 31	10, 13, 28, 38	14, 15, 18, 40	3, 36, 37, 10	29, 35 , 3, 5	2, 14, 30, 40	35, 28, 31, 40	28, 27, 3, 18
24		2, 22	26, 32	-	-	-	-	-	10
25	2, 5, 34, 10	35, 16, 32, 18		10, 37, 36,5	37, 36, 4	4, 10, 34, 17	35, 3, 22, 5	29, 3, 28, 18	20, 10, 28, 18
26	15, 20, 29	-	35, 29, 34, 28	35, 14, 3	10, 36, 14, 3	35, 14	15, 2, 17, 40	14, 35, 34, 10	3, 35, 10, 40
27	3, 10, 14, 24	2, 35, 24	21, 35, 11, 28	8, 28, 10, 3	10, 24, 35, 19	35, 1, 16, 11	-	11, 28	2, 35 3, 25
28	32, 13, 6	-	28, 13, 32, 24	32, 2	6, 28, 32	6, 28, 32	32, 35, 13	28, 6, 32	28, 6, 32
29	32, 28, 2	25, 10, 35	10, 28, 32	28, 19, 34, 36	3, 35	32, 30, 40	30, 18	3, 27	3, 27, 40
30	22, 23, 37, 35	34, 39, 19, 27	21, 22, 35, 28	13, 35, 39, 18	22, 2, 37	22, 1, 3, 35	35, 24, 30, 18	18, 35, 37, 1	22, 15, 33, 28
31	17, 2, 40	30, 18, 35, 4	35, 28, 3, 23	35, 28, 1, 40	2, 33, 27, 18	35, 1	35, 40, 27, 39	15, 35, 22, 2	15, 22, 33, 31
32	13, 29, 1, 40	35	35, 13, 8, 1	35, 12	35, 19, 1, 37	1, 28, 13, 27	11, 13, 1	1, 3, 10, 32	27, 1, 4
33	1, 16, 35, 15	4, 18, 39, 31	18, 13, 34	28, 13 35	2, 32, 12	15, 34, 29, 28	32, 35, 30	32, 40, 3, 28	29, 3, 8, 25
34	25, 2, 35, 11	1	34, 9	1, 11, 10	13	1, 13, 2, 4	2, 35	11, 1, 2, 9	11, 29, 28, 27
35	15, 35, 29	-	35, 10, 14	15, 17, 20	35, 16	15, 37, 1, 8	35, 30, 14	35, 3, 32, 6	13, 1, 35
36	34, 26, 6	1, 16	34, 10, 28	26, 16	19, 1, 35	29, 13, 28, 15	2, 22, 17, 19	2, 13, 28	10, 4, 28, 15
37	29, 1, 4, 16	2, 18, 26, 31	3, 4, 16, 35	36, 28, 40, 19	35, 36, 37, 32	27, 13, 1, 39	11, 22, 39, 30	27, 3, 15, 28	19, 29, 39, 25
38	35, 13, 16	-	28, 10	2, 35	13, 35	15, 32, 1, 13	18, 1	25, 13	6, 9
39	2, 6, 34, 10	35, 37, 10, 2	-	28, 15, 10, 36	10, 37, 14	14, 10, 34, 40	35, 3, 22, 39	29, 28, 10, 18	35, 10, 2, 18

293

	Duration of action of stationary object	Temperature	Illumination intensity	Use of energy by moving object	Use of energy by stationary object	Power	Loss of energy	Loss of substance	Loss of information
	16	17	18	19	20	21	22	23	24
1	-	6, 29, 4, 38	19, 1, 32	35, 12, 34, 31	-	12, 36, 18, 31	6, 2, 34, 19	5, 35, 3, 31	10, 24, 35
2	2, 27, 19, 6	28, 19, 32, 22	19, 32, 35	-	18, 19, 28, 1	15, 19, 18, 22	18, 19, 28, 15	5, 8, 13, 30	10, 15, 35
3	-	10, 15, 19	32	8, 35, 24	-	1, 35	7, 2, 35, 39	4, 29, 23, 10	1, 24
4	1, 40, 35	3, 35, 38, 18	3, 25	-	-	12, 8	6, 28	10, 28, 24, 35	24, 26,
5	-	2, 15, 16	15, 32, 19, 13	19, 32	-	19, 10, 32, 18	15, 17, 30, 26	10, 35, 2, 39	30, 26
6	2, 10, 19, 30	35, 39, 38	-	-	-	17, 32	17, 7, 30	10, 14, 18, 39	30, 16
7	-	34, 39, 10, 18	2, 13, 10	35	-	35, 6, 13, 18	7, 15, 13, 16	36, 39, 34, 10	2, 22
8	35, 34, 38	35, 6, 4	-	-	-	30, 6	-	10, 39, 35, 34	-
9	-	28, 30, 36, 2	10, 13, 19	8, 15, 35, 38	-	19, 35, 38, 2	14, 20, 19, 35	10, 13, 28, 38	13, 26
10	-	35, 10, 21	-	19, 17, 10	1, 16, 36, 37	19, 35, 18, 37	14, 15	8, 35, 40, 5	-
11	-	35, 39, 19, 2	-	14, 24, 10, 37	-	10, 35, 14	2, 36, 25	10, 36, 3, 37	-
12	-	22, 14, 19, 32	13, 15, 32	2, 6, 34, 14	-	4, 6, 2	14	35, 29, 3, 5	-
13	39, 3, 35, 23	35, 1, 32	32, 3, 27, 15	13, 19	27, 4, 29, 18	32, 35, 27, 31	14, 2, 39, 6	2, 14, 30, 40	-
14	-	30, 10, 40	35, 19	19, 35, 10	35	10, 26, 35, 28	35	35, 28, 31, 40	-
15	-	19, 35, 39	2, 19, 4, 35	28, 6, 35, 18	-	19, 10, 35, 38	-	28, 27, 3, 18	10
16	+	19, 18, 36, 40	-	-	-	16	-	27, 16, 18, 38	10
17	19, 18, 36, 40	+	32, 30, 21, 16	19, 15, 3, 17	-	2, 14, 17, 25	21, 17, 35, 38	21, 36, 29, 31	-
18	-	32, 35, 19	+	32, 1, 19	32, 35, 1, 15	32	13, 16, 1, 6	13, 1	1, 6
19	-	19, 24, 3, 14	2, 15, 19	+	-	6, 19, 37, 18	12, 22, 15, 24	35, 24, 18, 5	-
20	-	-	19, 2, 35, 32	-	+	-	-	28, 27, 18, 31	-
21	16	2, 14, 17, 25	16, 6, 19	16, 6, 19, 37	-	+	10, 35, 38	28, 27, 18, 38	10, 19
22	-	19, 38, 7	1, 13, 32, 15	-	-	3, 38	+	35, 27, 2, 37	19, 10
23	27, 16, 18, 38	21, 36, 39, 31	1, 6, 13	35, 18, 24, 5	28, 27, 12, 31	28, 27, 18, 38	35, 27, 2, 31	+	-
24	10	-	19	-	-	10, 19	19, 10	-	+
25	28, 20, 10, 16	35, 29, 21, 18	1, 19, 26, 17	35, 38, 19, 18	1	35, 20, 10, 6	10, 5, 18, 32	35, 18, 10, 39	24, 26, 28, 32
26	3, 35, 31	3, 17, 39		34, 29, 16, 18	3, 35, 31	35	7, 18, 25	6, 3, 10, 24	24, 28, 35
27	34, 27, 6, 40	3, 35, 10	11, 32, 13	21, 11, 27, 19	36, 23	21, 11, 26, 31	10, 11, 35	10, 35, 29, 39	10, 28
28	10, 26, 24	6, 19, 28, 24	6, 1, 32	3, 6, 32	-	3, 6, 32	26, 32, 27	10, 16, 31, 28	-
29	-	19, 26	3, 32	32, 2	-	32, 2	13, 32, 2	35, 31, 10, 24	-
30	17, 1, 40, 33	22, 33, 35, 2	1, 19, 32, 13	1, 24, 6, 27	10, 2, 22, 37	19, 22, 31, 2	21, 22, 35, 2	33, 22, 19, 40	22, 10, 2
31	21, 39, 16, 22	22, 35, 2, 24	19, 24, 39, 32	2, 35, 6	19, 22, 18	2, 35, 18	21, 35, 2, 22	10, 1, 34	10, 21, 29
32	35, 16	27, 26, 18	28, 24, 27, 1	28, 26, 27, 1	1, 4	27, 1, 12, 24	19, 35	15, 34, 33	32, 24, 18, 16
33	1, 16, 25	26, 27, 13	13, 17, 1, 24	1, 13, 24	-	35, 34, 2, 10	2, 19, 13	28, 32, 2, 24	4, 10, 27, 22
34	1	4, 10	15, 1, 13	15, 1, 28, 16	-	15, 10, 32, 2	15, 1, 32, 19	2, 35, 34, 27	-
35	2, 16	27, 2, 3, 35	6, 22, 26, 1	19, 35, 29, 13	-	19, 1, 29	18, 15, 1	15, 10, 2, 13	-
36	-	2, 17, 13	24, 17, 13	27, 2, 29, 28	-	20, 19, 30, 34	10, 35, 13, 2	35, 10, 28, 29	-
37	25, 34, 6, 35	3, 27, 35, 16	2, 24, 26	35, 38	19, 35, 16	19, 1, 16, 10	35, 3, 15, 19	1, 18, 10, 24	35, 33, 27, 22
38	-	26, 2, 19	8, 32, 19	2, 32, 13	-	28, 2, 27	23, 28	35, 10, 18, 5	35, 33
39	20, 10, 16, 38	35, 21, 28, 10	26, 17, 19, 1	35, 10, 38, 19	1	35, 20, 10	28, 10, 29, 35	28, 10, 35, 23	13, 15, 23

294

	Loss of time	Quantity of substance	Reliability	Measurement accuracy	Manufacturing precision	Object-affected harmful factors	Object-generated harmful factors	Ease of manufacture	Ease of operation
	25	26	27	28	29	30	31	32	33
1	10, 35, 20, 28	3, 26, 18, 31	3, 11, 1, 27	28, 27, 35, 26	28, 35, 26, 18	22, 21, 18, 27	22, 35, 31, 39	27, 28, 1, 36	35, 3, 2, 24
2	10, 20, 35, 26	19, 6, 18, 26	10, 28, 8, 3	18, 26, 28	10, 1, 35, 17	2, 19, 22, 37	35, 22, 1, 39	28, 1, 9	6, 13, 1, 32
3	15, 2, 29	29, 35	10, 14, 29, 40	28, 32, 4	10, 28, 29, 37	1, 15, 17, 24	17, 15	1, 29, 17	15, 29, 35, 4
4	30, 29, 14	-	15, 29, 28	32, 28, 3	2, 32, 10	1, 18	-	15, 17, 27	2, 25
5	26, 4	29, 30, 6, 13	29, 9	26, 28, 32, 3	2, 32	22, 33, 28, 1	17, 2, 18, 39	13, 1, 26, 24	15, 17, 13, 16
6	10, 35, 4, 18	2, 18, 40, 4	32, 35, 40, 4	26, 28, 32, 3	2, 29, 18, 36	27, 2, 39, 35	22, 1, 40	40, 16	16, 4
7	2, 6, 34, 10	29, 30, 7	14, 1, 40, 11	25, 26, 28	25, 28, 2, 16	22, 21, 27, 35	17, 2, 40, 1	29, 1, 40	15, 13, 30, 12
8	35, 16, 32 18	35, 3	2, 35, 16	-	35, 10, 25	34, 39, 19, 27	30, 18, 35, 4	35	-
9	-	10, 19, 29, 38	11, 35, 27, 28	28, 32, 1, 24	10, 28, 32, 25	1, 28, 35, 23	2, 24, 35, 21	35, 13, 8, 1	32, 28, 13, 12
10	10, 37, 36	14, 29, 18, 36	3, 35, 13, 21	35, 10, 23, 24	28, 29, 37, 36	1, 35, 40, 18	13, 3, 36, 24	15, 37, 18, 1	1, 28, 3, 25
11	37, 36, 4	10, 14, 36	10, 13, 19, 35	6, 28, 25	3, 35	22, 2, 37	2, 33, 27, 18	1, 35, 16	11
12	14, 10, 34, 17	36, 22	10, 40, 16	28, 32, 1	32, 30, 40	22, 1, 2, 35	35, 1	1, 32, 17, 28	32, 15, 26
13	35, 27	15, 32, 35	-	13	18	35, 24, 30, 18	35, 40, 27, 39	35, 19	32, 35, 30
14	29, 3, 28, 10	29, 10, 27	11, 3	3, 27, 16	3, 27	18, 35, 37, 1	15, 35, 22, 2	11, 3, 10, 32	32, 40, 28, 2
15	20, 10, 28, 18	3, 35, 10, 40	11, 2, 13	3	3, 27, 16, 40	22, 15, 33, 28	21, 39, 16, 22	27, 1, 4	12, 27
16	28, 20, 10, 16	3, 35, 31	34, 27, 6, 40	10, 26, 24	-	17, 1, 40, 33	22	35, 10	1
17	35, 28, 21, 18	3, 17, 30, 39	19, 35, , 3, 10	32, 19, 24	24	22, 33, 35, 2	22, 35, 2, 24	26, 27	26, 27
18	19, 1, 26, 17	1, 19		11, 15, 32	3, 32	15, 19	35, 19, 32, 39	19, 35, 28, 26	28, 26, 19
19	35, 38, 19, 18	34, 23, 16, 18	19, 21, 11, 27	3, 1, 32		1, 35, 6, 27	2, 35, 6	28, 26, 30	19, 35
20	-	3, 35, 31	10, 36, 23	-	-	10, 2, 22, 37	19, 22, 18	1, 4	-
21	35, 20, 10, 6	4, 34, 19	19, 24, 26, 31	32, 15, 2	32, 2	19, 22, 31, 2	2, 35, 18	26, 10, 34	26, 35, 10
22	10, 18, 32, 7	7, 18, 25	11, 10, 35	32		21, 22, 35, 2	21, 35, 2, 22		35, 32, 1
23	15, 18, 35, 10	6, 3, 10, 24	10, 29, 39, 35	16, 34, 31, 28	35, 10, 24, 31	33, 22, 30, 40	10, 1, 34, 29	15, 34, 33	32, 28, 2, 24
24	24, 26, 28, 32	24, 28, 35	10, 28, 23	-		22, 10, 1	10, 21, 22	32	27, 22
25	+	35, 38, 18, 16	10, 30, 4	24, 34, 28, 32	24, 26, 28, 18	35, 18, 34	35, 22, 18, 39	35, 28, 34, 4	4, 28, 10, 34
26	35, 38, 18, 16	+	18, 3, 28, 40	3, 2, 28	33, 30	35, 33, 29, 31	3, 35, 40, 39	29, 1, 35, 27	35, 29, 25, 10
27	10, 30, 4	21, 28, 40, 3	+	32, 3, 11, 23	11, 32, 1	27, 35, 2, 40	35, 2, 40, 26		27, 17, 40
28	24, 34, 28, 32	2, 6, 32	5, 11, 1, 23	+	-	28, 24, 22, 26	3, 33, 39, 10	6, 35, 25, 18	1, 13, 17, 34
29	32, 26, 28, 18	32, 30	11, 32, 1	-	+	26, 28, 10, 36	4, 17, 34, 26		1, 32, 35, 23
30	35, 18, 34	35, 33, 29, 31	27, 24, 2, 40	28, 33, 23, 26	26, 28, 10, 18	+	-	24, 35, 2	2, 25, 28, 39
31	1, 22	3, 24, 39, 1	24, 2, 40, 39	3, 33, 26	4, 17, 34, 26	-	+	-	-
32	35, 28, 34, 4	35, 23, 1, 24	-	1, 35, 12, 18	-	24, 2	-	+	2, 5, 13, 16
33	4, 28, 10, 34	12, 35	17, 27, 8, 40	25, 13, 2, 34	1, 32, 35, 23	2, 25, 28, 39	-	2, 5, 12	+
34	32, 1, 10, 25	2, 28, 10, 25	11, 10, 1, 16	10, 2, 13	25, 10	35, 10, 2, 16	-	1, 35, 11, 10	1, 12, 26, 15
35	35, 28	3, 35, 15	35, 13, 8, 24	35, 5, 1, 10	-	35, 11, 32, 31	-	1, 13, 31	15, 34, 1, 16
36	6, 29	13, 3, 27, 10	13, 35, 1	2, 26, 10, 34	26, 24, 32	22, 19, 29, 40	19, 1	27, 26, 1, 13	27, 9, 26, 24
37	18, 28, 32, 9	3, 27, 29, 18	27, 40, 28, 8	26, 24, 32, 28	-	22, 19, 29, 28	2, 21	5, 28, 11, 29	2, 5
38	24, 28, 35, 30	35, 13	11, 27, 32	28, 26, 10, 34	28, 26, 18, 23	2, 33	2	1, 26, 13	1, 12, 34, 3
39	-	35, 38	1, 35, 10, 38	1, 10, 34, 28	18, 10, 32, 1	22, 35, 13, 24	35, 22, 18, 39	35, 28, 2, 24	1, 28, 7, 19

	Ease of repair	Adaptability or versatility	Device complexity	Complexity of control	Extent of automation	Productivity
	34	35	36	37	38	39
1	2, 27, 28, 11	29, 5, 15, 8	26, 30, 36, 34	28, 29, 26, 32	26, 35 18, 19	35, 3, 24, 37
2	2, 27, 28, 11	19, 15, 29	1, 10, 26, 39	25, 28, 17, 15	2, 26, 35	1, 28, 15, 35
3	1, 28, 10	14, 15, 1, 16	1, 19, 26, 24	35, 1, 26, 24	17, 24, 26, 16	14, 4, 28, 29
4	3	1, 35	1, 26	26	-	30, 14, 7, 26
5	15, 13, 10, 1	15, 30	14, 1, 13	2, 36, 26, 18	14, 30, 28, 23	10, 26, 34, 2
6	16	15, 16	1, 18, 36	2, 35, 30, 18	23	10, 15, 17, 7
7	10	15, 29	26, 1	29, 26, 4	35, 34, 16, 24	10, 6, 2, 34
8	1	-	1, 31	2, 17, 26	-	35, 37, 10, 2
9	34, 2, 28, 27	15, 10, 26	10, 28, 4, 34	3, 34, 27, 16	10, 18	-
10	15, 1, 11	15, 17, 18, 20	26, 35, 10, 18	36, 37, 10, 19	2, 35	3, 28, 35, 37
11	2	35	19, 1, 35	2, 36, 37	35, 24	10, 14, 35, 37
12	2, 13, 1	1, 15, 29	16, 29, 1, 28	15, 13, 39	15, 1, 32	17, 26, 34, 10
13	2, 35, 10, 16	35, 30, 34, 2	2, 35, 22, 26	35, 22, 39, 23	1, 8, 35	23, 35, 40, 3
14	27, 11, 3	15, 3, 32	2, 13, 25, 28	27, 3, 15, 40	15	29, 35, 10, 14
15	29, 10, 27	1, 35, 13	10, 4, 29, 15	19, 29, 39, 35	6, 10	35, 17, 14, 19
16	1	2	-	25, 34, 6, 35	1	20, 10, 16, 38
17	4, 10, 16	2, 18, 27	2, 17, 16	3, 27, 35, 31	26, 2, 19, 16	15, 28, 35
18	15, 17, 13, 16	15, 1, 19	6, 32, 13	32, 15	2, 26, 10	2, 25, 16
19	1, 15, 17, 28	15, 17, 13, 16	2, 29, 27, 28	35, 38	32, 2	12, 28, 35
20	-	-	-	19, 35, 16, 25	-	1, 6
21	35, 2, 10, 34	19, 17, 34	20, 19, 30, 34	19, 35, 16	28, 2, 17	28, 35, 34
22	2, 19		7, 23	35, 3, 15, 23	2	28, 10, 29, 35
23	2, 35, 34, 27	15, 10, 2	35, 10, 28, 24	35, 18, 10, 13	35, 10, 18	28, 35, 10, 23
24	-	-	-	35, 33	35	13, 23, 15
25	32, 1, 10	35, 28	6, 29	18, 28, 32, 10	24, 28, 35, 30	-
26	2, 32, 10, 25	15, 3, 29	3, 13, 27, 10	3, 27, 29, 18	8, 35	13, 29, 3, 27
27	1, 11	13, 35, 8, 24	13, 35, 1	27, 40, 28	11, 13, 27	1, 35, 29, 38
28	1, 32, 13, 11	13, 35, 2	27, 35, 10, 34	26, 24, 32, 28	28, 2, 10, 34	10, 34, 28, 32
29	25, 10	-	26, 2, 18	-	26, 28, 18, 23	10, 18, 32, 39
30	35, 10, 2	35, 11, 22, 31	22, 19, 29, 40	22, 19, 29, 40	33, 3, 34	22, 35, 13, 24
31	-	-	19, 1, 31	2, 21, 27, 1	2	22, 35, 18, 39
32	35, 1, 11, 9	2, 13, 15	27, 26, 1	6, 28, 11, 1	8, 28, 1	35, 1, 10, 28
33	12, 26, 1, 32	15, 34, 1, 16	32, 26, 12, 17	-	1, 34, 12, 3	15, 1, 28
34	+	7, 1, 4, 16	35, 1, 13, 11	-	34, 35, 7, 13	1, 32, 10
35	1, 16, 7, 4	+	15, 29, 37, 28	1	27, 34, 35	35, 28, 6, 37
36	1, 13	29, 15, 28, 37	+	15, 10, 37, 28	15, 1, 24	12, 17, 28
37	12, 26	1, 15	15, 10, 37, 28	+	34, 21	35, 18
38	1, 35, 13	27, 4, 1, 35	15, 24, 10	34, 27, 25	+	5, 12, 35, 26
39	1, 32, 10, 25	1, 35, 28, 37	12, 17, 28, 24	35, 18, 27, 2	5, 12, 35, 26	+

296

Appendix 3. The 40 Inventive Principles with Examples

Each principle represents a concept or idea that may be applied to solve the problem situation. The 40 principles are the general concepts that Altshuller identified that inventors use to create new inventions. They are the principles used to remove a technical contradiction (according to TRIZ, an invention removes a contradiction). The principle may be applied to individual parts, the system as a whole, individual operations or steps, the functions performed, etc. When solving technical contradictions, two or several principles may be combined. For example use segmentation with mechanical vibration.

1. **Segmentation**
   - Divide into independent parts
     - *Break a process into multiple stages; use a sprinkler system instead of one hose.*
   - Divide an object into separate parts to make it easy to assemble or disassemble
     - *Modular "flat pack' furniture; sectional fishing pole.*
   - Increase the degree of segmentation or fragmentation, this may include transition to the micro level; use of particles, atoms, molecules, droplets, grains etc.
     - *Venetian blinds or plantation shutters instead of solid window blind.*
     - *Use charged particles to move air instead of a fan with blades.*
2. **Taking Out** (Extraction, separation)
   - Separate and remove the undesired property or part
     - *Locate a noisy air conditioning unit outside a home; use lead shielding to absorb harmful X-rays.*
   - Extract only the necessary or desired part or property
     - *Use the recorded sound of barking dogs to scare burglars.*
3. **Local Quality**
   - Change from uniform to non-uniform (structure or environment)
     - *A "slow lane" and "fast lane" on a highway; one wide door for wheelchair access.*
   - Have different parts perform different functions
     - *A pencil with eraser.*
   - Make the local conditions suitable for the functioning of each part
     - *Hot iron for applying heat, cool handle for holding; a clean (sterile) operating theater for surgery, normal environment for recovery.*
4. **Asymmetry** (Symmetry change)
   - Change symmetric to asymmetric
     - *Use wedge shapes, stir "off center" to create more mixing, asymmetric electrode sizes used in reactive ion etching to create enhanced ion acceleration.*
   - Increase degree of asymmetry
     - *Increase the aspect ratio (height to width ratio) to increase packing density or strength.*
5. **Merging** (Combining, integration, consolidation)
   - Merge or join together similar objects or related objects in space
     - *Twin hulled catamaran for stability, integrate/combine many radio telescopes to gain more power, use both sides of a circuit board.*
   - Merge or join together similar objects or related objects in space (perform operations in parallel if possible).
     - *Cut and collect grass in a lawnmower.*
6. **Universality** (Multi-functionality)
   - Perform multiple functions, eliminate the need for other parts or systems

        o   *Use a pencil as a ruler by adding measurement marks to it; a jar is also used as a cup.*

7. **Nested Doll** (Russian dolls)
   - Place one object inside another
     - o *Stacking cups to save space, add insulation to a pipe.*
   - Pass one object through another
     - o *a retracting antenna.*

8. **Anti-weight** (Counterweight)
   - To compensate for weight, merge with other elements or objects to provide lift
     - o *Blimp (balloon) to lift advertising sign, logs to float a raft.*
   - To compensate for weight, use aerodynamic, hydrodynamic, buoyancy and other forces.
     - o *Airflow to lift hovercraft, hydrofoils to lift (force) boat out of water.*

9. **Preliminary Anti-Action** (Prior counter action)
   - If an action produces both useful and harmful effects, perform an action in advance to reduce or eliminate the harmful effect
     - o *Hanging a weight on wire causes it to bend downwards, pre-bend the cable upwards so when the weight is placed on the wire, it will bend to the desired straight and level position.*
     - o *Use masking tape to stop undesired deposition while spray painting.*
   - Introduce stresses to oppose known undesirable stresses later
     - o *Steel bars are used to reinforce concrete, by pre-stressing the steel before pouring concrete a stronger structure can be obtained.*

10. **Preliminary Action** (prior action – prepare in advance)
    - Prepare in advance to perform or help perform a useful action
      - o *Pre-pasted wallpaper. Pre-drilled holes for easy assembly of modular furniture.*
    - Stage objects in advance so they can be applied from the most convenient location
      - o *Surgical instruments laid out in a specific order to be easily found. Fire extinguishers at visible locations. Prepare a problem solving flowchart.*

11. **Beforehand Cushioning** (Prior compensation)
    - For low reliability (or high risk) objects or systems, compensate in advance for a failure.
      - o *Seatbelt and airbags in cars. Safety barrier between opposite lanes of a road, first aid kit, fire extinguisher.*

12. **Equipotentiality** (Remove tension, bring to the same level)
    - Eliminate an undesired potential field or the need to work against a potential field
      - o *Gravity is a potential field; eliminate the need to work against gravity by rolling rather than lifting and carrying. Use locks in a canal to raise and lower boats.*
      - o *Eliminate static charge build-up using conductive grounding straps.*

13. **The other way round** (Do in reverse, inversion)
    - Perform the opposite action
      - o *Instead of heating, cool. To find pinholes in a film, instead of illuminating and inspecting the surface, illuminate from behind to easily show the holes.*
    - Change what moves to be fixed and what is fixed to have movement
      - o *Rotate the part instead of the tool. Move the cloth instead of needle and thread (sewing machine).*
    - Turn it upside down
      - o *Invert a partly used can of paint, the paint will seal the out the air.*

14. **Spheroidality** (Curvature)
    - Change from straight to curved, from flat to spherical, from cube to ball shaped
      - *Corrugated sheets for strength.*
    - Use rollers, balls, spirals, domes
      - *Ballpoint pen.*
    - Change motion from linear to rotary, use centrifugal forces
      - *Use ball casters instead of cylindrical wheels for greater maneuverability, spin dry.*
15. **Dynamics** (Dynamicity, dynamization, change parameters in time)
    - Perform changes in order to optimize for each stage of operation
      - *Adjustable anything, car seat, screwdriver that has many different head sizes, etc.*
    - If an object is fixed, make it have free motion
      - *Flexible screwdriver for using at corners, inflatable boat.*
    - Divide into parts capable of moving relative to each other
      - *Hinged ruler, truck and trailer, telescopic pointer, (also nested doll), chain.*
16. **Partial or Excessive Action** (Excess or shortage)
    - If the exact amount (100%) is difficult to achieve, use slightly more or slightly less
      - *When laying concrete, overfill then remove the excess, spray paint over masking tape then remove the tape when painting (also local quality).*
      - *Test only a sample, use a book light to illuminate only the pages of a book, use gold plating instead of solid gold.*
17. **Another Dimension** (Change to another dimension)
    - Move from one to two or three dimensional space
      - *2D movie to 3D movies.*
    - Stack objects on top of each other, side by side arrays
      - *Multi story buildings, multi-core cables, subway underground transport.*
    - Use another side
      - *Use a reflector to direct light in a flash-light; use both sides of a circuit board.*
18. **Mechanical Vibration** (Oscillation)
    - Use oscillations
      - *Vibrating toothbrush.*
    - Increase frequency
      - *Ultrasonic cleaning.*
    - Use resonance
      - *Microwave drying of water.*
    - Replace mechanical with piezoelectric vibrations
      - *Quartz watch.*
    - Use ultrasonic vibrations combined with electromagnetic fields (combine oscillating fields)
19. **Periodic Action**
    - Instead of a continuous action, replace it with a periodic (impulse) action
      - *Use many hammer blows to insert a nail, flashing lights on emergency vehicle, hop instead of run.*
    - If an action is periodic, change its magnitude or frequency
      - *Distinctive siren alternating between louder and lower frequency. FM radio.*
    - Use pauses between actions to perform other actions
      - *Carry out repairs when batteries are recharging.*

20. **Continuity of Useful Action**
   - Continuously work without a break, all parts work at maximum output
     - *Ballpoint pen has roller for continuous action (continuously supplies ink), hybrid car stores mechanical energy as electrical energy when breaking. The stored electrical energy is used to power the car to start moving again.*
   - Remove all idle and intermediate (in between) actions
     - *A ship carries a cargo of bananas in one direction. It returns loaded with oil.*

21. **Skipping** (Hurrying, rushing through)
   - Conduct a dangerous or harmful action at very high speed.
     - *Swallow medicine in a series of small quick sips to minimize the bad taste, firewalkers run across hot coals before the heat conducts to burn their feet.*

22. **Blessing in Disguise** (Turn harm to benefit, turn lemons into lemonade, make the problem into a good thing)
   - Use harmful factors to create a positive effect (especially surroundings or environment)
     - *Recycle waste, use leaking steam from a pipe to heat a greenhouse.*
   - Remove a harmful factor by combining it with another harmful factor
     - *Remove spilled oil by burning it off.*
   - Increase a harmful action to a level that it is no longer harmful (make better by making worse)
     - *Start a backfire to control the boundary of a forest fire. Take allergens to build up immunity to them.*

23. **Feedback** (Checking back)
   - Introduce feedback
     - *Automatic control, a floating ball valve controls the depth of water; statistical quality control.*

24. **Intermediary** (Insert a mediator)
   - Use an intermediary
     - *Glue to hold parts together, oil to reduce friction, telephone system, spring to buffer impact*
   - Introduce a temporary intermediary
     - *Oven gloves to carry hot objects, masking tape for spray painting, funnel to direct liquid when pouring from one vessel to another*

25. **Self Service**
   - An object performs useful maintenance or supplementary functions by itself
     - *Self sharpening knife, the blade is sharpened when moving into and out of its sheath, self winding watch (uses motion to charge), self service restaurant.*
   - Make use of waste resources
     - *Waste heat is converted to power.*

26. **Copying**
   - Use an inexpensive or simple copy in place of an original that is expensive, fragile or difficult to operate
     - *Flight simulator instead of using an aircraft for initial flying lessons, use crash test dummies (models); make a sacrificial copy for destructive measurement.*
   - Replace an object with an optical copy
     - *Use a photograph from which the information can be analyzed*
   - If an optical copy is used, replace with infrared or ultraviolet
     - *Infrared night vision camera.*

27. **Cheap Disposables** (Cheap short living objects)

- Replace an expensive object with a cheap one (or many cheap ones)
    - *Plastic cups, disposable gloves, syringes (contamination risk), Band-aid.*

28. **Mechanical Substitution**
    - Replace a mechanical system with a sensory (optical, acoustical, thermal or olfactory) system
        - *Use loud noise to clear an area, use the change in the pitch of sound of a drill to tell when drilling is complete, use signs rather than physically informing, smell to detect a gas leak, colored (dye) liquid to detect a leak.*
    - Use an electric, magnetic or electromagnetic field to interact with an object
        - *Use an electromagnet to lift metal.*
    - Replace fields that are stationary with mobile, static to variable, unstructured to structured
        - *A light bulb emits a constant light field, a switch makes it mobile turns on and off, a dimmer makes it variable. Mix liquids by placing a magnet into the mixture and rotating an external magnetic field, radio controlled aircraft.*
    - Use fields in conjunction with ferromagnetic particles or objects.
        - *Magnetically levitated platforms for moving parts through a factory.*

29. **Pneumatics and Hydraulics**
    - Use gases and liquids instead of solid parts
        - *Inflatable dingy, air cooling, pneumatic tire, air blower for cleaning, airbag in car, hydraulic lifting.*

30. **Flexible Shells and Thin Films**
    - Use flexible shells and thin films instead of solid structures
        - *Soft contact lenses instead of eyeglasses, car cover, flexible jar lid (indicates if previously opened), thin film of gold plating.*
    - Separate an object from its environment using flexible films or thin films
        - *Glass for an electric light bulb, paint, varnish, sealing tape, plastic bag, air gap for electrical insulation, vacuum gap (space) in dual pane windows for thermal insulation.*

31. **Porous Materials** (Holes)
    - Make an object porous or add porous elements
        - *Use holes as a filter to separate or pass materials, use holes or bubbles to: make an object lighter, reduce quantity of material used, increase volume, make softer or flexible, reduce conductivity, make absorbent, perform capillary action, perforations to make paper easy to tear.*
    - If already porous, fill holes with a substance
        - *In solid soap, fill holes with grit to make abrasive. Fill holes with lubricant that is released as the material erodes.*

32. **Color Changes** (change optical properties)
    - Change the color of an object or its surroundings
        - *Make a telephone box red to be easily seen, red light in a photography darkroom, yellow life vest.*
    - Change the transparency of an object or its surroundings
        - *Transparent syringe for checking dose, glass bottomed boat, glass refrigerator door.*
    - Use colored additives or luminescent additives
        - *Add colored dye to detect a leak, use luminous paint to see watch hands in the dark.*

302

33. **Uniformity** (Homogeneity)
- For a given object, make the other objects that interact with it from the same materials or from material with similar properties.
  - Stir water with a rod of ice, make the container that holds food edible, in semiconductor devices deposit adjacent thin films with similar coefficients of thermal expansion.

34. **Discarding and Recovering**
- Discarding: make objects that have completed their useful function disappear, evaporate, dissolve, etc.
  - Remove masking tape after painting, capsule for delivering medicine dissolves, biodegradable grocery bags.
- Recovering: restore consumables during operation
  - Self sharpening blades, water re-purification.

35. **Parameter Changes** (Transformation of properties)
- Change the Physical state (solid, liquid, or gas)
  - Transport propane gas in liquid form, cook vegetables in steam.
- Change concentration or density
  - Use a gel that is easier to control the spreading.
- Change degree of flexibility
  - Use more flexible material for vibration dampening.
- Change temperature
  - Implement thermal property changes that change chemical characteristics, viscosity, taste, preservation, etc.

36. **Phase Transition**
- Make use of the phenomena that occurs during a phase change (a release or absorption of energy, increase in volume, etc.)
  - Release of pressurized liquid gas can be used to cool, low temperatures to "dry" air and remove water molecules, expansion of water to steam to drive an engine, use crystals formed during solidification, shape memory alloys as actuators.

37. **Thermal Expansion** (Relative change)
- Use thermal expansion or contraction of materials
  - Fit a cold and hot pipe together. There will be a tight joint when ambient temperature is reached as they expand and contract.
- Use multiple materials with different coefficients of thermal expansion
  - Bi-metallic valve two sandwiched layers of differing metals which expand and contract at different lengths depending on temperature. This causes the bimetallic element to deflect or curve. It's connected to a valve which then controls the flow of a fluid.

38. **Strong Oxidants** (Enriched atmosphere, accelerated oxidation)
- Use oxygen enriched atmosphere
  - O2 in scuba diving, Air enhances bacterial growth in sewage plant .
- Use pure oxygen
- Expose air or oxygen to ionizing radiation (ionized oxygen)
- Use ozonized oxygen

39. **Inert Atmosphere** (Inert environment)
- Replace a normal environment with an inert one
  - Use inert gases. Store parts in a nitrogen environment to stop corrosion, argon gas in a light bulb, use filtered air in an operating theater, use foam to extinguish fire, use helium instead of a hydrogen balloon, plastic bags to store food.
- Add neutral substances or parts to an object or system
  - Nitrogen bubbles to agitate water, add neutral filler material to soap.

303

40. **Composite Materials** (Composite structures)
- Change homogeneous (uniform) materials to composite (made up of many parts or layers). The new composite materials enable many new products to be created because previously they were not available.
  - o *Fiberglass (glass fibers in a resin matrix) is very light and strong used to make canoes, surf boards, ceramic heat resistant tiles on the space shuttle, boron fiber material for a bullet proof vest, etc.*

Appendix 4. The 39 Parameter Definitions

1. **Weight of Moving Object**
The mass of or gravitational force exerted by a moving or mobile object. This also includes where there is any motion or mobility between parts related to the problem.
Commonly used terms: burden, load, heaviness.

2. **Weight of Stationary Object**
The mass of or gravitational force exerted by a stationary object.
Commonly used terms: burden, load, heaviness.

3. **Length (or Angle) of Moving Object**
Any linear or angular dimension relating to a moving or mobile object.
Commonly used terms: width, height, depth.

4. **Length (or Angle) of Stationary Object**
Any linear or angular dimension relating to a stationary object.
Commonly used terms: width, height, depth.

5. **Area of a Moving Object**
Any dimension related to surface area of a moving object.
Commonly used terms: region, space, zone.

6. **Area of Stationary Object**
Any dimension related to surface area of a stationary object.
Commonly used terms: region, space, zone.

7. **Volume of Moving Object**
Anything related to the three dimensional measure of space occupied by a moving object or the space around it.
Commonly used terms: capacity, space, size.

8. **Volume of Stationary Object**
Anything related to the three dimensional measure of space occupied by a stationary object or the space around it.
Commonly used terms: capacity, space, size.

9. **Speed**
The velocity or speed of an object or the rate of any kind of process or action.
Commonly used terms: pace, rate, rapidity.

10. **Force** (Intensity)
The capacity to physically change an object's condition.
Commonly used terms: force, pull, twist, torque, lift.

11. **Stress or Pressure** (Tension)
Force exercised on a unit area. Stress is the effect of forces on an object. Stresses can be tensile or compressive, static or dynamic.
Commonly used terms: compression, tension, elasticity, force.

12. **Shape**
The internal or external contour or appearance of a component or system
Commonly used terms: pattern, profile, curvature, outline, perimeter, edge, outside.

13. **Stability of the Object's Composition**
The resistance to change of the integrity of a system.
Commonly used terms: inertness, stability, robustness.

14. **Strength**
The ability to resist changing in response to force, the ability to absorb force. Resistance to breaking.
Commonly used terms: hardness, rigidity, robustness.

15. **Duration of Action by Moving Object**
The time that a moving object or system takes to perform a function or action.
Commonly used terms:: rate, period, frequency, time between failures, operating time.

16. **Duration of Action by Stationary Object**
The time that a stationary object or system takes to perform a function or action.
Commonly used terms: rate, period, time between failures, operating time

17. **Temperature**
Measured thermal condition of an object or system.
Commonly used terms: loss or gain of heat, conduction, convection, radiation.

18. **Illumination Intensity** (Brightness)
Light flux per unit area, color, brightness, light quality, etc.
Commonly used terms: reflectivity, emissivity, color.

19. **Use of Energy by Moving Object**
The amount of the energy spent. The actual amount of energy (not the efficiency of its use, which is captured by parameter 27 – Reliability).
Commonly used terms: consumption, heat-input, calories.

20. **Use of Energy by Stationary Object**
The amount of the energy spent. The same meaning as parameter 19, except there is no motion related to the problem .
Commonly used terms: consumption, heat-input, calories.

21. **Power**
The rate at which work is performed. The rate of use of energy. Rate of energy output.
Commonly used terms: action intensity, energy per unit time, wattage.

22. **Loss of Energy**
Waste of energy, input energy does not return a useful function.
Commonly used terms: inefficiency, loss of output.

23. **Loss of Substance**
Waste of substances, materials, subsystems, products, fields, etc.

305

Commonly used terms: erode, leak, desorption.

24. Loss of Information
Loss or waste of data or information.
Commonly used terms: misunderstanding, distortion, corruption.

25. Loss of Time
Waste of time – waiting periods, slack time, etc.
Commonly used terms: delay, duplication of effort, time lost on redundant or unnecessary activities.

26. Quantity of Substance
The amount or number of materials, substances, parts, fields or subsystems, etc. used.
Commonly used terms: things, parts, components.

27. Reliability (Robustness)
The ability to perform in a predictable way without degradation.
Commonly used terms: repeatability, Mean-Time-Between-Failure (MTBF), durability.

28. Measurement Accuracy
Degree of precision or accuracy, the degree to which a measurement is close to the actual value.
Commonly used terms: tolerance, error, standard deviation.

29. Manufacturing Precision (Consistency)
The degree to which the actual characteristics of a system or object correspond to the designed or required accuracy .
Commonly used terms: manufacturing accuracy, tolerance.

30. Object-Affected Harmful Factors
Any harmful factors or phenomena acting on a system or object that produces a harmful effect to reduce quality or effectiveness.
Commonly used terms: undesired effect, contamination, damage.

31. Object-Generated Harmful Factors
Harmful side-effects created by an object or system.
Commonly used terms: contamination, pollution, erosion, side effect, repercussion, secondary problem.

32. Ease of Manufacture (Manufacturability)
How simple or convenient it is to fabricate or produce a product.
Commonly used terms: convenience of assembly, design-for-manufacturability, easy build

33. Ease of Operation (Ease of Use)
Convenience of use, how easy something is to learn and operate.
Commonly used terms: user-friendly, learning-curve, familiarization time.

34. Ease of Repair (Repairability)
Convenience or simplicity with which a system is repaired and restored to its operating condition
Commonly used terms: ease of maintenance, assembly and disassembly, cleanability.

35. Adaptability or Versatility
The ability to respond to external changes.
Commonly used terms: flexibility of purpose, customizability.

36. Device Complexity

The quantity or diversity of system elements and their interactions or difficulty of use.
Commonly used terms: intricacy of use, high number of components, elaborateness of use.

37. **Complexity of Control**
The level of difficulty to monitor and control a system or object, the complexity of detection and measurement.
Commonly used terms: difficulty of monitoring or inspection, difficulty of measurement or detection.

38. **Extent of Automation**
The ability of a system or object to perform its functions without human interaction.
Commonly used terms: automatic, self controlled, intelligent.

39. **Productivity**
The amount of work or useful functions performed per unit of time.
Commonly used terms: work rate, output rate, value, operations.

Appendix 5. The Inventive Standards Flowchart

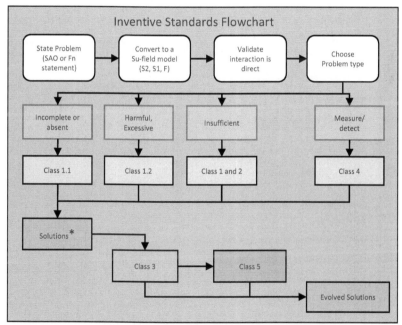

* Solutions are created by using Class 1.1, 1.2, 1, 2 and 4 to form an effective complete Su-field or an effective measurement and detection Su-field. The effective Su-field or "system" may be further developed by Class 3 and made more ideal by Class 5 inventive standards resulting in more evolved solutions.

Appendix 6. The 76 Inventive Standards

The 76 inventive standards are organized into 5 classes. Each class has groups of standards. They are numbered using a system of 1, 1.1, 1.1.1 etc. We have also included numbers 1-76 in parenthesis. Details and examples of the individual standards in each Class/Group is preceded by a summary for the class and group.

Class 1 Standards for Creation (synthesis) and Destruction of a Su-field

Class 1: is broken into two groups.
<u>Summary</u>
Class 1 Group 1
Class 1.1 deals with the creation of a Su-field (addressing an incomplete Su-field) and the improvement of an insufficient Su-field.
- A substance or field needs to be added to a Su-field to make it work or become effective.
    - o Complete the Su-field by introducing the missing substance or field or both.
    - o Introduce additives (substances/objects) to make Su-field work (temporary, internal, external/between or into the environment).
    - o Maximize/optimize the quantity of a substance or field that is used to make it more effective.
    - o Perform local intensification or weakening of the interaction to make it more effective.

Class 1 Group 2
Class 1.2 deals with how to address a harmful Su-field by the destruction of a Su-field interaction.
- The Su-field interaction should be destroyed or neutralized.
    - o Introduce a new substance between S2 and S1 to block the harmful interaction.
    - o Introduce a substance that's a modification of S1 or S2 to block or neutralize the harmful interaction by using a modified existing substance - modification is a more ideal solution than introducing a new substance.
    - o Add a third sacrificial substance that will draw the harmful effect.
    - o Introduce a second field to counteract the first (double Su-field).
    - o Turn off a magnetic field.

1.1.1 (1) An interaction needs two substances and a field to function. Either a substance or field is **absent**. We need to add a substance or a field to make the Su-field work or become effective **(Su-field Synthesis)**.

Example: A bottle contains a liquid, and suspended in the liquid are heavy solid particles. We wish to separate the particles from the liquid in order to pour off and collect some clean particle free liquid.

Incomplete Su-field	Complete Su-field (Effective)

Liquid holds particles. There is no field to separate the particles from the liquid. The Su-field model is incomplete. The flowchart directs us initially to 1.1. A field should be applied to complete the Su-field. Review the list of fields to help think of field types. The idea created by this is to apply centrifugal force by spinning the bottle to separate the particles from the liquid. Notice the substance S3 that supplies the field is not identified. In this case it's a person. The Su-field is simply there to help direct our thinking to what is needed to solve the problem.

1.1.2 (2) Introduce additives (S3) temporarily or permanently into S1 or S2 (Internal complex Su-field).

Example: Paint in an aerosol can separates into a dense and non-dense liquid when it has been stored for several days. Before use, the can is shaken for several minutes but inconsistent spray still occurs due to the insufficiently mixed paint. Problem: make the paint uniform so it will spray evenly.

When the can is shaken it insufficiently agitates the paint inside. The mechanical field between the can and paint is insufficient.

Insufficient Su-field   Internal Complex Su-field

Person moves (shakes) can. Can moves (shakes) paint. S2 is can, S1 is paint.
One solution prompted by 1.1.2 is to add some solid heavy beads (S3) into the can. This will improve the insufficient agitation. Note that you could search scientific effects for the function of how to agitate, stir, mix, etc.

Other examples:
- Add a dye to a liquid to help find a leak. Eye insufficiently detects liquid (via a light field), add a dye to make it easier to see.
- Take antibiotic to remove infection. Blood removes infection is incomplete (blood does not remove infection). Add a biological field and substance S2 (antibiotic).

1.1.3 (3) Introduce additives (S3) temporarily or permanently into S1 or S2, or between S1 and S2. (External Complex Su-field)

Example: A nut is used to seal the end of a threaded pipe to stop liquid flow, but often the surfaces of the nut and pipe do not seal properly and there is a leak.

Insufficient Su-field   External Complex Su-field

The idea is to make the interaction effective by adding an additive to the nut by placing a rubber ring over it to eliminate the space between the nut and pipe.

Other examples:
- Add a lubricant onto the surface of metal to reduce friction.
- Cool a drill bit with water.

Note, our problem could also be considered a harmful interaction. We don't need to use only class 1.2 for harmful interactions. The flowchart is a guide.

1.1.4 (4) Add a temporary or permanent additive from the environment into or onto (between) S1 and S2. This includes any locally available materials or resources including natural environmental resources (internal and external complex Su-fields).
Examples:
- Use snow to stop heat loss by building an igloo (air cools person).

- Use the melted wax of a candle to hold a candle in place.
- Sea water as ballast to level a cargo ship.

1.1.5 (5) Change or modify the environment (may use additives, fields, parameters etc.). Note this does not have a standard basic Su-field diagram to help explain.
Examples :
- If you feel cold, add a heater to heat the air.
- Humidify a room to stop skin drying.
- Chemical 1 and 2 don't react. Heat them, and a thermally driven reaction takes place.

1.1.6 (6) For a system (interaction) that requires precise quantities of substances, but is difficult to measure the exact amount, **minimum mode** is used to provide the minimum or optimized amount. Use the maximum amount and remove the excess.
Examples:
- Apply excess paint coating and remove (scrape or spinoff or drip off) the excess.
- Apply excess photoresist and spin to achieve the required thickness.
- Gold plated jewelry instead of solid gold.

1.1.7. (7) Use excess field and remove the excess.
For a system that requires precise quantities but the action cannot be reduced, use **maximum mode**. Apply the excess field and use an intermediary link (use a substance or field) to reduce it to the required level.
Examples:
- Melt chocolate in a bath (second saucepan) within a pot of water, the temperature produced by the thermal field will not exceed 100C.
- Use a lampshade to reduce the brightness of a bulb.

1.1.8.1 (8a) Selective Maximum Mode (maximum field), the field is strong and cannot be reduced, use a protective substance S3 as a mask to eliminate or drop the intensity in the desired areas.
Examples:
- Sunglasses to protect eyes.
- Photomask in semiconductor manufacture.

1.1.8.2 (8b) Selective Maximum Mode (minimal Field). Apply the weak field to specific locations to improve the interaction.

Example:
- Place explosives at specific locations to maximize their impact.

Class 1 Group 2

1.2.1 (9) There is a useful and harmful action between S2 and S1, and S2 does not have to be in contact with S1: introduce a new (foreign) substance to block the harmful action.

| Interaction contains harmful Su-field | Add S3 between S2 and S1 |

Examples:
- Sun heats person but also damages skin. Use sunscreen gel to block the harmful ultraviolet rays.
- Teflon to stop food sticking to a pan also allows heat to transfer for cooking.

1.2.2 (10) There is a useful and harmful action between S2 and S1, and S2 does not have to be in contact with S1, BUT new substances cannot be added. Introduce a modified existing substance to block the harmful action.

F1

S2   S1
Interaction contains Harmful
Su-field

F1

S2   S1
S3*
Add S3* between S2 and S1

S3* is a modified S2 or S1.

A field or fields may be added to modify S2 or S1. Solutions include adding "nothing" by using holes, gaps, bubbles, foam, etc.

Examples:
- A pipe is used to transport fast moving water (useful). Water turbulence in a pipe causes damage to the internal surface of the pipe (harmful). Solution: freeze the outside of the pipe to form a layer of ice on the inside, the ice will protect the pipe from turbulence damage.
- A cloth canopy is used to stop rain but it is easily blown by the wind (harmful). Add vents to neutralize the force of the wind on the umbrella.

Note that when a field is added, something is used to create it that is not included in the Su-field model. We used a substance that can refrigerate water to freeze it.

1.2.3 (11) Introduce a sacrificial substance (S3) to absorb or "draw off" the effect of a harmful action.

F1

S2   S1

Interaction contains
harmful Su-field

F1

S2   S1
S3
Add S3 between S2 and S1

Examples:
- A lightning rod to absorb electrical charge.
- Use a heat sink to absorb thermal energy to stop the effect of overheating.
- Insulate pipes to stop cracking due to thermal effects.

1.2.4 (12) There is a useful and harmful action between S2 and S1 but unlike 1.2.1 and 1.2.2, S1 and S2 **do** have to be in contact. 1.2.4 is to introduce a second field F2 to counteract the harmful field F1 (here we use a **Double Su-field** to illustrate the solution).

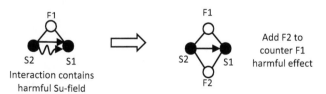

F1

S2   S1
Interaction contains
harmful Su-field

F1

S2   S1

F2

Add F2 to
counter F1
harmful effect

In removing or neutralizing the harmful field the new field F2 may also add a new useful action.

Examples:
- A mechanical field (friction) opposes a platform sliding along a pathway. Use a counteracting magnetic field to levitate the platform.
- Cardboard packages are damaged by friction as they slide down a ramp. Add a vibrational field to allow the packages to flow freely without frictional damage.
- Knife is used to perform surgery but also produces pain. Use a chemical field (anesthetic) to neutralize pain.

Note that when a field is added, something (a substance) is used to create it and it is not included in the Su-field model. For the cardboard packages example above, we used a substance (vibration generator) to create vibration F2, but the substance used to create the vibrational field F2 is not shown, only the field is shown. This is the convention in Su-field modeling.

1.2.5 (13) Relates to counteracting a harmful magnetic field. A magnetic field can be removed by heating the magnetic material above the Curie point or by using an opposite magnetic field.

Examples:
- Use an electromagnet to turn off the magnetic field when lifting objects containing ferro-magnetic materials.
- Use the Curie point to release ferro-magnetic materials above a specific temperature.

Class 2 Standards for Development of the Su-field Model

Class 2: is broken into four groups.
Summary
Class 2 inventive standards strengthen and develop an insufficient interaction.

2.1 Deals with an insufficient **field**. The field is weak so move to a chain or double Su-field to "boost" the interaction.
- A chain Su-field is made by adding a field F2 and introducing a new substance S3 in series (next to S2 or S1). The new field enhances the existing one.
- A double Su-field is created by adding a second field in parallel (between S2 and S1) that compensates for the weak field. S3, the object that supplies F2 in parallel, by convention is not shown in the double Su-field model notation.

2.2. Also deals with an insufficient field, in this case by trying to develop the resources. Use the internal resources more effectively to improve functionality (without introducing new fields and substances and by using internal resources more). The field is weak, try using the internal resources more.
- Use a more evolved field (refer to the MATCHEM mnemonic)
- Transition to micro level
- Use porous or capillary materials
- Dynamize (joint-chain-elastic-liquid-gas-field)
- Increase the coordination of rhythm of substances and fields

2.3 Tries to improve an insufficient interaction by redistributing energy more effectively by co-coordinating rhythm, re-distributing energy more effectively, using changes in frequency:
- Use frequency: continuous actions, pulse, constant frequency, resonance
- Structure fields and substances
- Coordinate frequency, rhythm, timing of fields

2.4 Ferromagnetic solutions are used to try to improve an insufficient interaction. Integrating ferromagnetic magnetic fields and magnetic materials is an effective way to improve the performance of a system. Altshuller had a specific interest in ferromagnetic materials. It is often suggested that too much emphasis is placed on ferromagnetism.

Class 2 Group 1

2.1.1(14) Add a substance S3, in-between or next to S2 or S1 (often S3 is a type of engine).

Insufficient Su-field   Chain Su-field

Examples:

- Gravity is used to move water from a hot water tank through the pipes of a building to heat it but it moves too slowly and water is cold towards the end of the pipe. Add a pump to increase the flow of water.
- To fit a nut to a bolt, using fingers alone provides insufficient torque; add a wrench (S3).

2.1.2 (15) Form a double Su-field by adding a second field to boost the field between S1 and S2. This normally requires using a third substance S3 that can produce the field F2; S3 not normally shown.

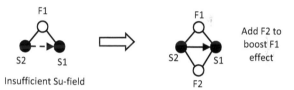

Insufficient Su-field

Examples:

- Parts are cleaned in a water bath. Add vibrations (vibrational mechanical field) to enhance the cleaning process.
- Screws are inserted using a mechanical screwdriver. Use an electrically powered screwdriver to enhance effectiveness.

Class 2 Group 2

2.2.1 (16) Use a more evolved field which offers improved control. Instead of gravity use mechanical field. Instead of mechanical use chemical etc, the use of the mnemonic (MATCHEM) may help.

Insufficient Su-field   Effective Su-field

Examples:

- Water is mechanically pumped up to a reservoir for future irrigation. Use an electrical pump instead.
- Metal parts are moved by a crane with a bucket. Use an electromagnet to pick them up instead.

2.2.2 (17) Transition of S2 to the micro level (S2 is the tool, the subject that acts on S1 the object). Think of micro level as the fragmented or segmented level of S2. If S2 is a rock, think of particles, grains, molecules, even liquid, gas, etc.

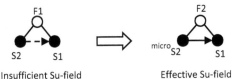

| Insufficient Su-field | Effective Su-field |

Examples:
- Use many individual "sprinklers" to water a lawn evenly instead of a single hose.
- "Sandblast" using a jet of grains of solid material to clean a surface instead of scraping.
- Ship fragile objects by surrounding with foam chips.

2.2.3 (18) Use porous or capillary materials*.

Use cavities, holes, capillaries, porous material to perform a function or add a substance.

Examples:
- Manufacture bricks with a hollow inside to improve insulation and reduce the amount of material.
- Shower head to distribute gas evenly.
- Perforations to filter or separate materials.
- Paper towels used to soak up liquid by capillary action.
- Add grains of solid disinfectant throughout slow dissolving gel. The delivery of disinfectant will remain constant over a long period.

*Capillary action is a weak force that allows a liquid to move up a thin tube (the narrower the tube the higher the rise) the force is driven by surface tension of the liquid. Porous materials will also draw liquids into them by a similar mechanism.

2.2.4 (19) Dynamize.

Add flexibility to the system (S1, S2 or F1), the ability to change parameters in time. Instead of fixed, use moving parts or fields, make them adjustable. See also the trend of dynamization; move from single joint, multiple joints, flexible materials, liquid, gas, plasma, etc.

Examples:
- Instead of a rigid pole for casting a fishing line, use a flexible pole.
- Use a flexible screwdriver for operating around corners.

2.2.5 (20) Structurize Fields

Temporarily or permanently organize fields such that their effect is controlled, use a specific frequency, use resonance frequency, use pulsed fields, standing waves, interference patterns, etc. to improve the effectiveness of an interaction.

Examples:
- Pulse pressure to aid drilling of rock.
- Polarized light to remove glare.

2.2.6 (21) Structurize Substances (objects). Temporarily or permanently change the structure so the interaction is more effective by moving to a mixed or evolving the spatial structure (1D to 2D to 3D).

Above we discussed creating a porous structure (2.2.3), we can also change irregular or homogenous (uniform structure or composition throughout) substances' effectiveness by moving to a mixed structure or by evolving the spatial structure (evolve from point to line to surface to volume) from 1D to 3D structure see trend of dynamization.

Examples:
- Add an internal steel framework to concrete to create re-enforced concrete.
- Instead of a fishing line, use a net.

Class 2 Group 3

2.3.1 (22) Co-ordinate the frequency of F1 to match or mismatch S2 or S1.

Use frequency to improve the interaction. Systems evolve from constant to gradient, impulse, resonance, interference. Improve the interaction by considering moving from a constant to a more evolved field. See trend of dynamization.

Examples:
- Heart massage at the frequency that the heart beats.
- Use the resonance frequency to disintegrate kidney stones.
- Mismatch the resonance frequency of a bridge with the frequency of steps created by people walking on it.

2.3.2 (23) Match (coordinate) rhythm of F1 and F2 in a chain or double Su-field to improve the effectiveness.

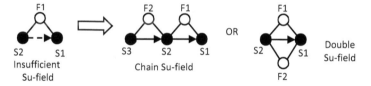

Insufficient
Su-field

Chain Su-field

Double
Su-field

- Fly fishing pole frequency adjusted to match length of line.
- Noise cancelling headphones F2 runs 180 degrees out of phase from F1.

2.3.3 (24) Coordinate independent fields (perform F1 during the downtime of F2).

Examples:
- Photocopier matches light exposure with placement of pages being copied.
- Clean car windscreen while filling car with gas.

Class 2 Group 4

2.4.1 (25) Add a ferromagnetic field and/or magnetic field to the system.

Example: Magnetic levitation

2.4.2 (26) Move to a more controlled field (or more evolved as in 2.2.1 but make it a magnetic field).

Combine 2.2.1 (move to more controlled or evolved fields) and 2.4.1 (use ferromagnetic fields and magnetic fields).

Example:
- Control the shape of rubber containing ferromagnetic material by applying a magnetic field.

315

2.4.3 (27) Use magnetic liquids

Use magnetic liquid. Magnetic liquids are a special case of 2.4.2. They are colloidal ferromagnetic particles suspended in kerosene, silicone or water.

Examples:
- Magnetorheological car suspension control, fluid inside piston contains ferromagnetic particles, an applied magnetic field can change viscosity controlling dampening of suspension.

2.4.4 (28) Apply capillary structures in ferromagnetic fields.

Example: A valve made up of ferromagnetic tube material between magnets. The alignment is controlled by magnetic fields.

2.4.5 (29) Use additives to give non-magnetic materials magnetic properties.

Use additives (such as a coating) to give non magnetic object magnetic properties (temporary or permanent).

Example: Add magnets or ferro magnetic materials into parts and move them by moving magnetic fields.

2.4.6 (30) Add ferromagnetic materials to the environment.

Introduce ferromagnetic materials into the environment if it's not possible to make the object magnetic.

Example: Use a magnetized tray to hold tools without magnetizing the tools themselves.

2.4.7 (31) Use effects (natural phenomena) of magnetic materials.

Example: Loss of magnetism above the Curie point can be used as a switch (materials lose magnetism when a specific temperature (the Curie point) is reached).

2.4.8 (32) Dynamize using magnetism.

Example: Electric motor.

2.4.9 (33) Structure using magnetism.

Example: Add ferromagnetic particles to a material then move the particles to change its structure by applying a magnetic field.

2.4.10 (34) Match rhythm of ferro-field models.

Example: Microwave ovens apply energy at the natural frequency water molecules absorb

2.4.11 (35) Use electric current to create magnetic fields instead of using magnetic particles.

Example: Electromagnets.

2.4.12 (36) Rheological liquids have their viscosity controlled by an electric field.

They can be used with any of the methods and mimic liquid-solid phase transitions.
Example: See 2.4.3, any situation where a liquid can be transitioned to a solid state.

Class 3: Standards for System Transition to the Supersystem and Micro Level

Class 3 is broken into two groups

Summary

In Class 1 and 2 we improved the interaction by adding to or modifying resources. Now we look for opportunities for improving functionality by developing the system. We look to the supersystem and micro levels.

Class 3 Group 1

3.1 Function moves to a supersystem. Conversion to bi-poly systems, we develop the system by transitioning to the supersystem. For example, instead of using coal power station, use solar power.

- Mono-bi-poly – simplest transition to the supersystem, (increase the functionality of the interaction to two or many *similar* substances).
- Evolve links; improve functionality by combining substances together.
- Functions become more diverse, improve functionality by linking different substances e.g. pencil with eraser.
- Trimming - Improve functionality by having substance (component) perform additional actions and removing the need for one or more substances (components).

Class 3 Group 2

3.2 Transition to micro level. Develop the system by transitioning its operation to the grain, particle, molecule, subatomic, etc, level.

Class 3 Group 1

3.1.1 (37) Transition to a bi or poly System.
 	
Follows the trend of mono-bi-poly. This adds more functionality by using multiple of the same S2, S1 or F1. It is the simplest conversion to incorporate external components (incorporate many of the same).
 	
Examples:
    • Single barreled shotgun, improve system functionality by using double barreled.
    • Cut many sheets of paper at a time instead of one.
    • Use a showerhead to spread water instead of a single nozzle.

3.1.2 (38) Evolve the links in bi and poly systems.
 	
Instead of many of the same things stuck together, evolve how they interact.
 	
Examples:
    • Catamaran can widen space between hulls depending on stability needed
    • Instead of many TV screens with same image, coordinate to make a large image each screen showing a part.

3.1.3 (39) Transition by increasing the difference between components of the bi- and poly systems. Incorporate additional functions. Example: • Pen becomes multicolored pen, adds a laser pointer, and adds a blade for cutting.
3.1.4 (40) Simplification (a form of trimming). Incorporate multi systems into one. Bi system, combine two chairs, poly, combine many chairs. Multi-purpose poly system, incorporate multi functions in one system instead of many different systems.

Example:
- Swiss army knife, mobile phone and camera, etc.

3.1.5 (41) Transition to incorporate the opposite features to the whole or the systems parts.

Examples:
- Pencil with eraser
- Claw hammer with claw for extracting nails.
- Bicycle chain parts are solid but flexible when linked.

Effective inventions often incorporate the ability to perform the anti-function of their main useful function.

Class 3 Group 2

3.2.1 (42) Transition to the micro level.
Develop the system by performing actions and using substances at the level of particles, grains, molecules, cells, bacteria, etc.

Examples:
- Semiconductor Industry electrical characteristics controlled at the level of electrons and atoms.
- Instead of sliding parts on rollers, use oil (extremely small parts).

Class 4: Standards for Measurement and Detection

Class 4 is broken into five groups.
Summary
Class 4 deals with how to improve or establish methods of measurement and detection.
4.1 Change the system so there is no need to measure (use roundabout or workaround methods). Measure copies or marks or labels to make measurement easier (e.g. barcode labels and scanner. If you cannot measure continuously, make discrete measurements).
4.2 Measure indirectly or measure additives or derivatives. Create a measurement system (some substances and fields need to be added to measure).
4.3 Enhance or strengthen the measurement system. Use resonance measurements.
4.4 Use ferro-magnetic fields (Curie point, etc.) Transition to ferromagnetic measurement Systems. Note that the focus on ferromagnetic measurement is somewhat dated. The introduction of ferromagnetic materials for measurement was included in the list before the development of other methods of remote sensing.
4.5 Evolve the measuring system .Transition to bi and poly systems (measure two or more at a time) or use first and second derivatives (measure speed and acceleration instead of distance).

Class 4 Group 1

4.1.1 (43) Eliminate the need to measure/detect.

Examples:
- To fill a tank to a specific level, put a hole in it at the desired level. Use thermocouples to control an air conditioning system to a specific temperature range.

4.1.2 (44) Measure a copy or an image. Introduce mark to make measurement or detection easier.

Examples:
- Take a photograph, if measurement is destructive, measure a sacrificial copy. Introduce a badge that can be scanned, barcodes.

4.1.3 (45) Use two separate detections instead of trying to measure.

Examples:
- To check if a weight is around 5 pounds. Put it on a scale that can check whether it is heavier than 4 and then check on another scale whether it is lighter than 6 pounds. It is shown to be between 4 and 6 pounds.

Class 4 Group 2

4.2.1 (46) Use a single or double Su-field with a field as the output. The output field correlated to what we are trying to measure or detect.

This simply means if there is no output field to measure or detect, create one by adding substances or fields. If the field is too weak then add substances or fields.

Examples (no field):
- A balloon has a very small hole that cannot be seen. Squeeze it and listen if air comes out. You added a substance (hand, and field mechanical pressure) to create a sound field which you detected.
- Tap to see if the inside is hollow.

Example (weak field):
- You cannot hear the sound of the heart clearly. Add a stethoscope (substance) to boost the sound.

4.2.2 (47) Measure the change in an additive.

Example:
- Add dye to the inner layer of a tire. When the dye color appears the outer rubber has eroded to its limit.

4.2.3 (48) Measure the change in an additive into the environment.

Example:
- Throw grass into the air to estimate wind speed and direction.

4.2.4 (49) Get the additive by decomposing (deconstructing) the environment.

Example:
- Disturb sediment to indicate the speed of water.

Class 4 Group 3

4.3.1 (50) Measure using scientific effects (correlate the change in the effect to the parameter to be measured).

Example:
- Measure the level of contamination of water by electrical conductivity.

4.3.2 (51) Measure resonance.

Example: Use resonance of a tuning fork to tune the tension of a piano wire.

4.3.3 (52) Measure the resonant frequency of an object joined or bound to the object you are trying to measure.

Example:
- Measure the resonance frequency of the gas in a chamber above a liquid to determine the quantity of liquid in the vessel.

Class 4 Group 4

4.4.1 (53) Transition to Ferro-Field models (Measuring Fe-Field).

Add or use ferromagnetic substances and a magnetic field (using permanent magnets or loops of current) to facilitate measurement.

Example:

- Detect cars by ferromagnetic components in the street (the timing of when to change traffic lights can be controlled by detecting cars).

4.4.2 (54) Add magnetic particles to a system or change a substance to facilitate measurement by detection of the magnetic field.

4.4.3 (55) If ferromagnetic particles can't be added directly into a system or substance construct a complex system; add ferromagnetic particles into that additional substance.

4.4.4 (56) Add ferromagnetic particles to the environment if they can't be added into the system.

4.4.5 (57) Measure the effect of natural phenomena associated with magnetism.

Examples:

- Curie point to measure temperature, the Hall effect, Magnetic Resonance Imaging (MRI).

Class 4 Group 5

4.5.1 (58) Transition to the bi and poly systems. If it's difficult to measure one, measure two or more. Combine measurements.

Example:

- The temperature of a termite is difficult to measure. Fill a jar with many termites and measure their temperature.

4.5.2 (59) Measure the first or second derivatives (in time or space).

Example:

- Instead of measuring position (distance), measure speed (1st derivative) the rate of change of distance or 2nd derivative (acceleration) the rate of change of velocity.

Class 5: Recommendations for more Ideal Application of the Standards

Class 5 is broken into five groups
Summary
Class 5 deals with how to best apply the inventive standards in Classes 1 through 4.

The most ideal way (minimum cost and maximum functionality) is suggested.
5.1 How to introduce or add a substance,
   o   Use workarounds/indirect ways
   o   Add "nothing," a void or hole instead of a substance
   o   Add a field instead
   o   Add external instead of internal additive
   o   Use small concentrated amounts
   o   Use temporary instead of permanent additives
   o   Use a copy or model
   o   Use something from the environment or something that disappears
   o   Use many small additives instead of one big one
   o   Use inflatables
   o   Use fields
5.3 Make use of the phases of substances, (gas instead of liquid, solid instead of gas, etc.). Use phase transitions, sometimes S2 and S1 are or contain substances (materials) for which we can change their phase, meaning, solid, liquid, gas, etc.
5.4 Application of Scientific Effects - use substances that control themselves, use amplification.
5.5 Use higher or lower forms of substances (e.g. electrolysis to release hydrogen from water).

Class 5 Group 1
5.1.1 (60) Workarounds (indirect ways of introducing a substance).
5.1.1.1. Add "nothing." Instead of adding a substance, add spaces, gaps, cavities, use air, bubbles, capillaries, holes, foam, etc. Example: • To reduce heat transfer, instead of adding a substance, use foam (add bubbles). Cavity wall insulation.
5.1.1.2 Instead of a substance, add a field. Example: • Instead of adding a latch on a door, use a magnet. Tap wood to find the hollow space inside instead of inserting a probe. Use a pressure field to determine if rubber gloves have a hole.
5.1.1.3 Instead of an internal additive use an external one. Example: • Instead of adding a control system into a robot, add it remotely. Instead of adding a dye to a liquid to help find a leak (eye insufficiently detects liquid via a light field) add a viscous liquid externally and look for bubbles. Adding externally is often simpler than internal addition.
5.1.1.4 Use a small amount of a very active substance. Example: Computer to control traffic rather than many traffic police.
5.1.1.5 Concentrate the additive at a very specific location. Example: • Spot of highly active detergent on a stain.
5.1.1.6 Introduce the additive temporarily. Example: Antibiotic to kill infection.
5.1.1.7 Use a copy or model (instead of the actual substance). Example: • Crash dummies. Videoconference.
5.1.1.8 Use a safe compound that reacts to produce the desired unsafe compound. Example: • Instead of adding chlorine (gas) to a swimming pool, add salt to a swimming pool, then by passing the saltwater through electrodes, chlorine is safely produced.
5.1.1.9 Obtain the additive by decomposition of a substance (S2 or S1) or the environment. Example: • Use waste as fertilizer.
5.1.2 (61) Divide elements into smaller units. Example: • Use two propellers instead of one very big one. Use many buckets instead of one big one.
5.1.3 (62) Additive eliminates itself after use (substances that disappear). Example: • Use sutures that disintegrate after a wound is healed. Dry ice pellets to "bead blast" clean a surface.

5.1.4 (63) Use inflatable structures or foam.

Example:
- Use an air bag to lift an object if a jack cannot be inserted.

Class 5 Group 2

5.2.1 (64) Use one field to create another field.

Example:
- Solar power (light field) is converted to electrical power via the photoelectric effect. Use heat to drive a chemical reaction (field). Microwave to heat cup of water.

5.2.2 (65) Use a field from the environment.

Example:
- Use air (wind) to dry clothes. Solar power to charge a battery, bees to mechanically distribute pollen. Use air to passivate the surface of aluminum to avoid corrosion.

5.2.3 (66) Use substance source fields.

Example:
- Radioactive pellets to destroy tumors. Add flowers to a room to smell nice (olfactory field).

Class 5 Group 3

5.3.1 (67) Substituting the phases.

Example:
- Transport gas as a liquid, convert back to a gas for use.

5.3.2 (68) Use dual phases (dynamic phases, the phase changes in time).

Example:
- Ice melts as a skate passes over then returns to ice.

5.3.3 (69) Use phase changes to perform an effect.

Example:
- Heat water to produce steam to drive an engine.

5.3.4 (70) Transition two phases (bi-phase system).

Example:
- Solid particles of sand in water for abrasive cleaning.

5.3.5 (71) Introduce an interaction (physical or chemical) between phases (created in 5.3.4) or parts of the system

Example:
- Instead of water, use acid to chemically remove the surface to aid polishing.

Class 5 Group 4

5.4.1 (72) Self controlled transitions.
The element of the system should change state by itself.

Example:
- Sunglasses that pass less light in bright sunlight. Smart materials that change shape by themselves.

5.4.2 (73) Amplify (optimize) the output field when there is a weak input field (often done near a transition point).

Example:
- Oxygen is stored in liquid form under high pressure. Use a regulator to lower the

pressure to breathe. Transistor amplifies radio signal information.

Class 5 Group 5
5.5.1 (74) Create substances at a micro level by decomposing. Example: •   Use electrolysis of water to release hydrogen and oxygen.
5.5.2 (75) Create particles or substances by combining smaller particles (of a simpler structural level). Example: •   Combine short chain polymers to create long chain polymer materials (plastics).
5.5.3 (76) How to apply 5.5.1 and 5.5.2 If a substance of high structural level has to be decomposed and it can't be decomposed, start with a substance of the next lower level structure. Similarly if a substance must be formed from materials of a low structural level and it cannot be, then start with the next higher level of structure.

76 inventive standards table contents derived and published with permission from materials published by Terninko, J., Miller, J., Domb, E., (2000) The Seventy-Six Standard Solutions, with Examples. February, March, April, June, July issues of www.TRIZ-journal.com

Appendix 7. Table of Specific Inventive Principles to Solve Physical Contradictions.

Inventive Principles useful for prompting ideas to solve physical contradictions
Satisfaction
Meet opposing requirements, often by the use of a smart material or chemical or scientific effect
36 phase transitions, 37 thermal expansion, 28 mechanical substitution/another sense, 35 parameter change, 38 strong oxidation and 39 inert atmosphere
Bypass
Use a totally different method to bypass the problem often by a scientific effect
25 self service, 6 multi functionality and 13 the other way round
Separation in Space
A characteristic must be present in one place but not present in another
1 segmentation, 2 taking out, 3 local quality, 17 another dimension, 13 other way round, 14 curvature, 7 nested doll, 30 flexible shells/thin films, 4 asymmetry, 24 intermediary and 26 copying
Separation in Time
A characteristic is present at one time and not present at another.
15 dynamization, 10 prior action, 19 periodic action, 11 beforehand cushioning, 16 partial or excessive action, 21 skipping, 26 copying, 18 mechanical vibration, 37 thermal expansion, 34 discarding and recovering, 9 prior counter action, and 20 continuity of useful action
Separation in Relation
A characteristic must be present for one action and not present for another.
13 the other way round, 35 parameter changes, 32 color changes, 36 phase transition, 31 porous materials, 38 strong oxidants, 39 inert atmosphere, 28 mechanical substitution and 29 pneumatics and hydraulics
Separation at System Level
A characteristic exists at the system level but not at the component level (or vice versa)
Transition to the subsystem (or micro-level)
1 segmentation, 25 self service, 27 cheap disposables, 40 composite materials, 33 uniformity and 12 equipotentiality
Transition to the supersystem
5 merging, 6 universality, 23 feedback and 22 blessing in disguise

Appendix 8. Flowcharts Roadmaps/Templates

8.1. *TRIZICS Roadmap*
8.2. *TRIZICS Problem Solving Template*
8.3. *Standard Structured Problem Solving Flowchart*
8.4. *Standard Structured Problem Solving Template*
8.5. *Standard structured Problem Solving incorporating TRIZ tools for Root Cause Analysis Flowchart*
8.6. *Standard Structured Problem Solving Incorporating TRIZ Tools for Root Cause Analysis Template*
8.7. *Contradiction Problem Solving Flowchart – Simplified ARIZ*
8.8. *Simplified ARIZ Algorithm*
8.9. *ARIZ-85C*

For all technical problems, use and start with and follow the TRIZICS Roadmap and template this is the **master roadmap and template**. The initial steps instruct us to apply the Standard Structured Problem Solving Flowchart (Appendix 8.3) and Template (Appendix 8.4) or if the root cause is unknown, apply the flowchart (Appendix 8.5) and template for Standard Structured Problem Solving incorporating TRIZ tools for Root Cause Analysis (Appendix 8.6). Root cause must be known before proceeding further.

If root cause is known, proceed to the next step (Step 2) of the TRIZICS Roadmap by selecting the problem type and applying the appropriate analytical tools which will produce a specific problem to solve. Reformulate the problem, use the instructions contained in the section "Which Solutions Tool (s) to Choose" see Chapter 1 Section 2.3 Step 5 as a guideline for which to choose. If a breakthrough inventive solution is sought, apply the basic tools for solving contradictions, A and B Sections of Simplified ARIZ (Appendix 8.7) using the Simplified ARIZ Algorithm (Appendix 8.8). If needed apply the advanced tools for solving contradictions, Sections C and D of Simplified ARIZ. If preferred, for a difficult to solve contradiction problem, use ARIZ 85-C (Appendix 8.9).

Appendix 8.1 TRIZICS Roadmap

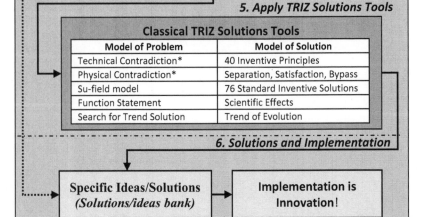

TRIZICS ROADMAP

| Problem |

1. Identify Problem

Standard Structured Problem Solving (see SSPS Algorithm)

Inventive Problem Solving - TRIZ

Ideas/ solutions created by Standard Structured Problem Solving,

2. Select Problem Type

Specific Problem (reactive or proactive)

General Inventive Goal (proactive)

Cause Unknown (1)

Cause known (2)

Improvement/ Development (3)

Failure Prevention (4)

3. Apply Analytical Tools

Root Cause Analysis

Root Cause Analysis incorporating TRIZ tools Algorithm
- Includes: CEC-1, SA-1

Analytical tools 2

Ideal Solution/ System, CEC-2 Functional Modeling and Trimming, Nine Windows, DTC Operator

Analytical tools 1

CEC-3 (why is system inadequate) S-curve analysis Trends of Evolution Anti-system

Subversion Analysis

Subversion Analysis (SA-2, CEC-4)

Ideas/ solutions created by finding root cause or by analytical tools.

4. Define Specific Problem

Define Specific Problem(s)

5. Apply TRIZ Solutions Tools

Classical TRIZ Solutions Tools

Model of Problem	Model of Solution
Technical Contradiction*	40 Inventive Principles
Physical Contradiction*	Separation, Satisfaction, Bypass
Su-field model	76 Standard Inventive Solutions
Function Statement	Scientific Effects
Search for Trend Solution	Trend of Evolution

6. Solutions and Implementation

Specific Ideas/Solutions *(Solutions/ideas bank)*

Implementation is Innovation!

326

*Technical and physical contradictions may also be addressed using the advanced tools contained in ARIZ (see Appendix 8.9 ARIZ-85C) and the Simplified ARIZ algorithm (see Appendices 8.7 and 8.8 and Chapter 9).

Appendix 8.2 TRIZICS Problem Solving Template

1.1 Problem Statement	

1.2 Apply the Standard Structured Problem Solving template If root cause is still unknown - Apply the Standard Structured Problem Solving incorporating TRIZ tools for Root Cause Analysis template	See separate flowcharts and templates 8.3/8.4 and 8.5/8.6

Complete the Standard Structured Problem Solving Template and Standard Structured Problem Solving Incorporating TRIZ Tools for Root Cause Analysis Template if it is needed.

1.3 Problem Statement (revised)	

Revise the problem-statement if it has been changed by performing Step 1.2

The cause of the problem must be understood before applying TRIZ tools.

TRIZICS Step 2 Select Problem Type

2.1 Problem type	

Select Type 2, 3 or 4 (Type 1 problems are not admissible, root cause must be known before Step 2)

Problem Types:
Specific Problem (may be reactive or proactive)

1. Solve a specific problem when the root cause is unknown.
2. Solve a specific problem for which the cause (mechanism of how it occurs) is known.

Proactive (General Inventive Goal)

3. Improve, develop, invent a technical system, or technical process.
4. Prevent future failures for a technical system or technical process.

Examples of the four problem types:

1. Eliminate intermittent leaking of a pipe in a gas supply system (specific, cause unknown).
2. Eliminate the fracturing of a glass tube due to thermal expansion when it is heated (specific, cause known).
3. Determine how to develop a motor car to gain market advantage. Invent a better floor cleaner. Improve the efficiency, repairability and quality of a preventative maintenance process (general inventive goal to improve a technical system or technical process).
4. Eliminate the causes of failure for a metal electroplating process, a radio, roller coaster, a kettle (failure prevention).

328

TRIZICS Step 3 Apply Analytical Tools

Analytical tools are chosen by problem type.

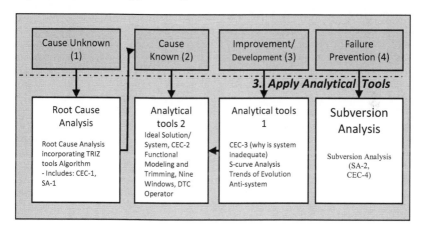

Analytical Tools For Type 2 Problems (Specific with known root cause)

Analytical Tools 2

Ideal Solution (Type 2)

Step 1 State the ideal solution, one where the problem solves itself. How can the system eliminate the problem by itself, using only the available resources of the system, surrounding environment and free or inexpensive resources?	
Step 2 Using the concept of the ideal solution, brainstorm ideas for how to solve the problem for free or using the systems local environment or inexpensive resources.	
Step 3 Using the list of ideas; create a list of solutions to solve the problem.	

CEC-2 (Type 2)

CEC-2 checks root cause, sometimes root cause is incorrectly identified and we are not working on the right problem or are working on a symptom rather than the cause.

Step 1: Write the problem in first box.	
Step 2: Continue building the CEC until causes are revealed.	
Step 3: Review the chain and create a list of the causes then create a plan to address these potential causes that includes solving the specific problems identified.	

Functional Modeling and Trimming (Type 2)

Step 1: Identify the system, subsystem, process, process step that contains the problem.	
Step 2 :Build a functional model of the technical system or technical process that contains the known problem. For a technical system build both the graphical and matrix models.	
Step 3 : Identify the specific problem interaction or problem step.	
Step 4: Can the problem be solved by trimming (removing a component or step and having it performed by another component or step).	

When the problem root cause is known it is important to try to include the problem as a function in the functional model. For example, if there is a leaking pipe, the pipe holds water insufficiently should be included in the model. If parts are being broken, include the function that causes it; arm transports part harmfully, etc.

Nine Windows (Type 2)

Step 1: Write the "problem" in center square.	
Step 2: Consider various timescales and levels of system assess whether the problem can be solved at that level/timeframe.	

DTC Operator (Type 2)

Step 1: Define problem: name the system or the part of system of interest.	
Step 2: Consider ideas created by DTC extremes: **Dimension:** If dimensions were extremely large, how could the problem be solved? If dimensions were extremely small, (almost gone) how could the problem be solved? **Time** If time were extremely long, how could the problem be solved? (E.g. days/years instead of seconds) (Or if speed were extremely slow what would change?) If time were extremely small, how could the problem be solved? (E.g. microseconds instead of seconds) (Or if speed were extremely fast what	

would change?) **Cost:** Not just in terms of dollars but cost in terms of downside, harmful effects, etc. If cost were extremely large, how could the problem be solved? If cost were extremely small, how could the problem be solved?	

Analytical Tools for a Type 3 Problem.

Analytical Tools 1 (Apply to Type 3 Problem)

CEC-3 (Type 3)	

Step 1: Start with "the technical system or technical process is not ideal."	
Step 2: Build the Cause-Effect Chain	
Step 3: Define specific problems to solve to improve the system	

S-curve Analysis (Type 3)	

- S-curve analysis is used to predict and manage development of a technical system along the S-curve.
  - o S-curve is used for planning future development of a product or process. Main parameters of a system or process can be evaluated for their position on the S-curve and development strategies created.

Step 1: Identify the main parameters of a technical system or technical process.	
Step 2: For each parameter, estimate its position on the S-curve (which stage it is at).	
Step 3: Define a plan to develop each parameter. Plan should identify specific problems.	

Trends of Evolution (Type 3)	

Step 1: Name the system or component . Note that a system may have several components each treated individually in terms of level of development for each trend.	
Step 2: On an arbitrary scale of 1-10, (1 is low, 10 is highly developed), estimate the level of development relative to each trend.	
Step 3: Define a plan for development of the system focused on improving the less developed features of the system to a more evolved state.	
Step 4: Repeat for each component of the system.	

The Anti-system (Type 3)

Step 1: Name the system and its main useful function.	
Step 2: Describe the anti-system or anti-action.	
Step 3: Create ideas, how would the anti-system work, what can it be used for?	

Analytical tools 2 are applied differently for a Type 3 (general) problem than a Type 2 (specific) problem.

For Type 3 Problems (General improvement of technical systems or technical processes)

Ideal System (Type 3)

Step 1: Describe the ideal system (or process Note that the ideal system is no system; the ideal process is no process the functionality is provided for free, no costs, and no resources would be used if the system were ideal.	
Step 2: Using the concept of the ideal system/process. Brainstorm ideas for how to improve the ideality of the system/process	
Step 3: Using the list of ideas of the improvements; create a list of solutions to improve the system or process.	

Functional Modeling and Trimming (Type 3)
Technical System

Step 1: Main Useful Function Statement
    MUF:
    Product:
For a specific problem, create a function statement that describes the problem, and then include the interaction in the functional model of the technical system or general system where the problem resides.	

Step 2: Make an inventory of all the components of the system. Main components are: C1, C2, C3 etc...	
Step 3:Make an inventory of the supersystem elements (external elements that are not designed as part of the system but can interact with it). Supersystem components are: SS1, SS2, SS3 etc.	
Step 4: Build a Graphical and Matrix Functional Model Functional Model of system (graphical) Functional model of system (matrix format)	
Step 5: Identify the interactions for improvement.	
Step 6: Create a trimming problem Solve trimming problem. How to make a trimmed system work?	

332

Technical Process

Step 1: Define the purpose of the procedure.	
Step 2: Define the main Phases (main blocks it can be broken into)	
Step 3: Define the individual steps and their quality	
Step 4: Make a list of the problem steps to improve	
Step 5 Define Improvement Plan including trimming Ideas a) Can individual steps be eliminated, merged or performed in parallel to reduce time/cost? b) Can we improve the sequence to be more ideal? c) What steps can be performed by another step? d) Define how to improve the problem steps.	

Nine Windows (Type 3)

Step 1: Write the "System" or "Problem" in center square.	
Step 2: Consider various timescales and levels of system. Assess whether the system can be improved or developed at that level/timeframe.	
Step 3: Consider ideas prompted by the anti-system (opposite function or action or state to the original nine windows).	

DTC Operator (Type 3)

Step 1: Define problem: name the system or the part of system of interest.	
Step 2: Consider ideas created by DTC extremes: **Dimension:** If dimensions were extremely large, how could the problem be solved/system developed? If dimensions were extremely small, (almost gone) how could the problem be solved/system developed? **Time:** If time were extremely long, how could the problem be solved/system developed? (E.g. days/years instead of	

333

seconds) Or if speed were extremely slow what would change? If time were extremely small, how could the problem be solved/system developed? (E.g. microseconds instead of seconds) Or if speed were extremely fast what would change? **Cost:** Not just in terms of dollars but cost in terms of downside, harmful effects, etc. If cost were extremely large, how could the problem be solved/system developed? If cost were extremely small, how could the problem be solved/system developed?	

Analytical tools to apply to a Type 4 Problem (failure prevention)

Subversion Analysis

Subversion Analysis SA-2 (CEC-4)	
Step 1: Identify the system	
Step 2: Information collection	
Step 3: Describe what can go wrong • Brainstorm a list of potential problems that can occur before parts and resources are used, during use and after use. o It is useful to perform a Cause-Effect Chain analysis (CEC-4) with the problem statement as "the system failed …." • Rank the identified potential failure modes in order of likelihood and severity of each problem and prioritize which problems to pursue first.	
Step 4: List the prioritized failure modes to be prevented and state inversely in the form "how to cause the problem" • How to cause problem No 1 • How to cause Problem No 2, etc.	

Play devil's advocate. Given all the resources, their storage, manufacture, the environment, people, procedures ,etc. how can you cause the problem? How to cause the problem in different circumstances, when idle, during use, etc? It is useful to task a group of content experts to provide a prioritized list of ways of how to cause the problem given all possible available resources.

334

Applying the analytical tools often leads to a re-definition of the original problem and the definition of a number of different new problems to address. For example, the starting problem may be the glass breaks, the specific problem is the how to stop vibration, or how to protect glass (in our next step we learn how to formulate these into the formats of the classical TRIZ solutions tools). These problems should be listed and kept track of in a "problems bank" where they can be prioritized.

Problems Bank		
1		Priority
2		
etc		

Formulate specific problems from the analysis above in one or more the "model of the problem" formats below. Prioritize forming a technical contradiction "if-then-but" statement if the goal is to find a breakthrough solution. If the problem is difficult to solve or a more innovative solution is needed, apply the tools of ARIZ.

Classical TRIZ Solutions Tools	
Model of Problem	**Model of Solution**
Technical Contradiction*	40 Inventive Principles
Physical Contradiction*	Separation, Satisfaction, Bypass
Su-field model	76 Standard Inventive Solutions
Function Statement	Scientific Effects
Search for Trend Solution	Trend of Evolution

*Technical and physical contradictions may also be addressed using the advanced tools contained ARIZ (see Appendix 8.9 ARIZ-85C) and the Simplified ARIZ algorithm (see Appendices 8.7 and 8.8 and Chapter 9).

A key skill for applying TRIZ is to formulate your specific problems into the model formats to be able to apply the TRIZ solutions tools.

Guideline for which tool to choose:
1. If the problem is simple and can be formulated as a basic interaction problem, form the problem into Su-field and function statements to apply the 76 inventive standards and search for scientific effects for solutions.
2. If a breakthrough is needed, form a technical contradiction (an "if-then-but" statement) and apply the basic tools for solving technical contradictions – the 40 inventive principles.
3. When there is a technical contradiction we can also form a physical contradiction and apply separation, satisfaction and bypass, assisted by scientific effects and a specific list of inventive standards.
4. If the problem is difficult to solve as a contradiction then apply the advanced tools for solving technical and physical contradictions. We took the key tools of ARIZ for solving technical and physical contradictions and named them the advanced tools. We created a flowchart to form and solve a contradiction which first applied the basic tools then the advanced tools (the tools of ARIZ). Our flowchart is named "Contradiction Problem Solving Flowchart - Simplified ARIZ" is shown in Section 2.5 of Chapter 9 Part 1 and Appendix 8.7.

5. ARIZ is used to formulate your problem into a contradiction and then solve it. ARIZ is recommended for difficult problems and has a reputation of being difficult to use, we recommend using simplified ARIZ discussed above as a simpler alternative.
6. To apply trends of evolution as a solutions tool, simply review the each trend and determine if any inspire a solution.

Model	Format
Technical Contradiction	If (state what change is made), then (state what good happens), but (state what bad happens). Example: If I increase power then speed improves but the shaft temperature rises.
Physical Contradiction	X-resource must be A to provide the useful action and anti-A to eliminate the negative action, or, X must be A and anti-A. Example: The cork must be solid and porous.
Su-field model	Subject-Action-Object statement where action is a field and subject and object are "things" but not a field. Example: Cork holds bottle (excessive tension/mechanical field).
Function Statement	Subject-Action-Object statement where action is a function and subject acts on the object. Example: Bearings heat shaft (harmful)
Search for a trend of evolution solution	Review the trends; sometimes they can prompt an idea for solving a specific problem.

TRIZICS Step 5 Apply TRIZ Solutions Tools

Su-field Model	Solve by Inventive Standards Flowchart and 76 Inventive Standards

Function Statement	Solve by finding methods to perform a function. Search scientific effects.

Search for a trend of evolution solution	Review the trends, search TRIZ body of work for additional trends.

Use Simplified ARIZ:

Use Basic Contradiction Problem Solving Tools

Apply Parts A and B of Simplified ARIZ. (see roadmap in Appendix 8.7 and template in Appendix 8.8)	

If needed, use Advanced Contradiction Problem Solving Tools

Apply Parts C and D of Simplified ARIZ. (see roadmap in Appendix 8.7 and template in Appendix 8.8)	

If needed Apply ARIZ-85C (see 8.9)	

TRIZICS Step 6 Solutions and Implementation

Solutions

Solutions Bank		
Source		Priority

Implementation Plan

	Action	Priority	Projected Completion Date	Owner	Status
1					
2					
3					
Etc					

337

Appendix 8.3 Standard Structure Problem Solving Flowchart

For any Problem the TRIZICS Roadmap leads us first to the Standard Structured Problem Solving Roadmap

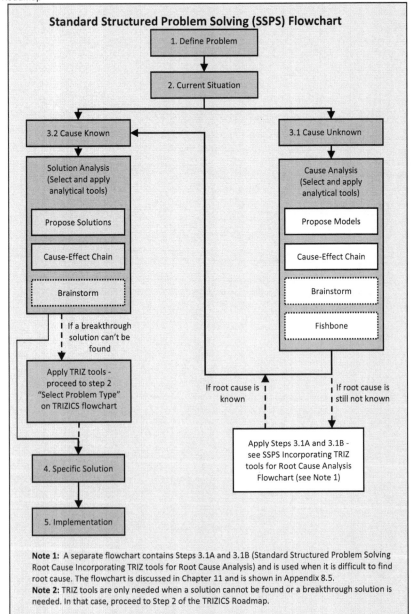

Standard Structured Problem Solving (SSPS) Flowchart

- 1. Define Problem
- 2. Current Situation
- 3.2 Cause Known
- 3.1 Cause Unknown
- Solution Analysis (Select and apply analytical tools)
- Cause Analysis (Select and apply analytical tools)
- Propose Solutions
- Propose Models
- Cause-Effect Chain
- Cause-Effect Chain
- Brainstorm
- Brainstorm
- Fishbone
- If a breakthrough solution can't be found
- Apply TRIZ tools - proceed to step 2 "Select Problem Type" on TRIZICS flowchart
- If root cause is known
- If root cause is still not known
- Apply Steps 3.1A and 3.1B - see SSPS Incorporating TRIZ tools for Root Cause Analysis Flowchart (see Note 1)
- 4. Specific Solution
- 5. Implementation

Note 1: A separate flowchart contains Steps 3.1A and 3.1B (Standard Structured Problem Solving Root Cause Incorporating TRIZ tools for Root Cause Analysis) and is used when it is difficult to find root cause. The flowchart is discussed in Chapter 11 and is shown in Appendix 8.5.

Note 2: TRIZ tools are only needed when a solution cannot be found or a breakthrough solution is needed. In that case, proceed to Step 2 of the TRIZICS Roadmap.

It is not necessary to use all of the analytical tools. It is recommended that the user first propose models/solutions then applies Cause-Effect Chain analysis. Use brainstorming and fishbone tools only if needed in addition to CEC analysis.

Appendix 8.4 Standard Structured Problem Solving Template

Step 1 Define Problem

1.1 Problem Statement	
Draw the problem Include pictures, photos, etc.	
1.2 Name the technical system or technical process	
1.3 What is the main useful function?	
1.4 What is the impact or cost of not solving the problem?	
1.5 What are the success criteria, the definition of "done"?	
1.6 What is the timeline for getting a solution?	
1.7 What are the limitations and requirements?	

Step 2 Current Situation

2.1 Containment Plan	

2.2 Problem Background/history Problem Background description Data/results	

2.3 Plans	

Action	Priority	Projected Completion Date	Owner	Status

2.4 Resources Needed	

Step 3 Analysis

3.1 Cause Unknown Analysis

Investigate Root Cause	

Provide details of experiments to be run, data to be collected etc.

Action	Purpose	Priority	Projected Completion Date	Owner	Status

339

From the description of the problem and the existing data listed in Step 2, list what key conclusions can be drawn from the data that support solving the problem.

	Data	Possible Conclusion	How data was validated
1	Pipes only leak after 3 months	Pipes are being slowly eroded	Analysis of recorded install dates.
2			
3			
Etc			

Check data and Conclusions

It is important to be vigilant
- Have new people check the information and provide fresh thinking
- Determine whether the conclusions can be wrong
- Assign a specific person (owner) responsible for checking the data
- Physically check/ witness the results rather than assuming information or data
- Always challenge calibration methods
- What potentially hidden resources might be present and how could they cause the problem.
- Describe a new or unusual mechanism that would have to exist to cause the problem.
- Demonstrate whether the problem is abnormal.

Describe Leading Model (s)

Leading Model(s) of the root cause

1	
2	
etc	

3.1.1 Propose Cause Models

A	
B	
C	
etc	

Standard Tools for Creative Thinking

3.1.2 Brainstorming

A	
B	
C	
etc	

A	
B	
etc	

3.1.4 CEC-1

A	
B	
etc	

3.2 Cause Known (Analysis)
Cause is known, step 3.2 is to identify solutions.

3.2.1 Propose Ideas

A	
B	
etc	

3.2.2 Brainstorm

A	
B	
etc	

3.2.1 CEC-2

A	
B	
etc	

If a solution can't be found, or a breakthrough solution is needed, use TRIZ. See TRIZICS Roadmap Step 2 – Select Problem Type.

Step 4 Ideas/Solutions

Solutions Bank		
Source		Priority

Step 5 Implementation Plan

	Action	Priority	Projected Completion Date	Owner	Status
1					
2					
3					
Etc					

Appendix 8.5 Standard Structured Problem Solving Incorporating TRIZ tools for Root Cause Analysis Flowchart

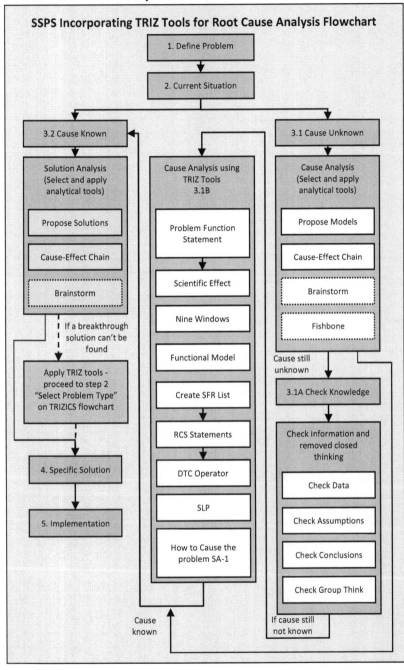

Note that it is not necessary to use all of the analytical tools and the user may choose in which sequence to use them.

Appendix 8.6 Standard Structured Problem Solving Incorporating TRIZ Tools for Root Cause Analysis Template

If root cause can't be found using the Standard Structured Problem Solving flowchart and template, use the Standard Structured Problem Solving incorporating TRIZ tools for Root Cause Analysis flowchart.

Step 1 Define Problem

1.1 Problem Statement	

Draw the problem
Include pictures, photos, etc.

1.2 Name the technical system or technical process	
1.3 What is the main useful function?	
1.4 What is the impact or cost of not solving the problem?	
1.5 What are the success criteria, the definition of "done"?	
1.6 What is the timeline for getting a solution?	
1.7 What are the limitations and requirements?	

Step 2 Current Situation

2.1 Containment Plan	

2.2 Problem Background/History	
Problem Background Description	
Data/results	

2.3 Plans	

	Action	Priority	Projected Completion Date	Owner	Status

2.4 Resources Needed	

343

Step 3 Analysis

3.1 Cause Unknown Analysis

Investigate Root Cause	

Provide details of experiments to be run, data to be collected etc.

	Action	Purpose	Priority	Projected Completion Date	Owner	Status

Summarize Conclusions	

From the description of the problem and the existing data listed in Step 2, list what key conclusions can be drawn from the data that support solving the problem.

	Data	Possible Conclusion	How data was validated
1	Pipes only leak after 3 months	Pipes are being slowly eroded	Analysis of recorded install dates.
2			
3			
Etc			

Check data and Conclusions	

It is important to be vigilant
- Have new people check the information and provide fresh thinking.
- Determine whether the conclusions can be wrong.
- Assign a specific person (owner) responsible for checking the data.
- Physically check/ witness the results rather than assuming information or data.
- Always challenge calibration methods.
- What potentially hidden resources might be present and how could they cause the problem?
- Describe a new or unusual mechanism that would have to exist to cause the problem.
- Demonstrate whether the problem is abnormal (or outlier).

Describe Leading Model (s)	

Leading Model(s) of the root cause

1	
2	
etc	

3.1.1 Propose Cause Models	

A	
B	
etc	

344

Standard Tools for Creative Thinking

3.1.2 Brainstorming	
A	
B	
etc	

3.1.3 Fishbone	
A	
B	
C	
etc	

3.1.4 CEC-1	
A	
B	
etc	

If Root Cause is Still Not Known (note these steps are additional to the standard problem solving template and are tools for helping determine root cause).

Step 3.1 A Check Knowledge

Check data and Conclusions	

Psychological inertia, erroneous data, wrong assumptions and false conclusions are often why root cause is not identified. It is important to keep an open mind and eliminate the causes of psychological inertia. It is important to validate and challenge all data and assumptions. It is necessary to prove the data being used is valid and to ensure the correct conclusions are drawn.

Below is a list of frequent reasons root cause is not identified and recommended actions the problem solver should take to address those issues.

Frequent reasons root cause is not identified:
- **Experience**: too much experience can lead to psychological inertia and drive thinking in a "trained" direction closing off new ideas.
- **Fixed thinking techniques**: repeating the same steps and using the same methods leads to repeating the same result, creating the same ideas.
- **Group think**: over time, a set of individuals working on a project will tend to think the same way, believe the same conclusions and results. This group mindset leads to psychological inertia, as new members are introduced instead of pursuing new ideas provided by "a fresh pair of eyes," the group tries to assimilate new members to existing thinking.
- **Model worship:** a specific "favorite" model is pursued and alternatives are dropped.
- **False information/incorrect data/false assumptions**: this is may be due to the way the data or information was collected. For example, incorrect calibration of measurement standards, or simply incorrect information or facts have been obtained or assumed.
- **False Conclusion**: for example, the sun rises every day in the east. False conclusion - the sun revolves around the earth.

- **Hidden resources**: the problem is caused by contaminants, or secondary resources, etc.
- **Hidden mechanism**: mechanism may be a new or unusual phenomenon or be an effect outside the problem solvers field of engineering or science.
- **There is no actual problem**: the result is "normal" simply a "rarity" or "outlier" of the normal distribution.
- **There is more than one problem** and therefore more than one cause.
- **Insufficient technical knowledge**: this is rarely the reason for a problem's root cause not being identified. Normally such gaps in knowledge are quickly closed and problems solved.

Actions to address common issues that impede root cause determination:
- Have new (different) people check all data and information to provide fresh thinking.
- Determine whether the conclusions can be wrong (be highly critical of all conclusions).
- Check the information is indisputable, assign a specific person (owner) responsible for checking the data.
- Physically check and visually witness information or data rather than accepting validation from others.
- Always challenge calibration methods and standards (hardware) used.
- Determine what potentially hidden or secondary resources might be present and how they could cause the problem.
- Describe a new or unusual mechanism that would have to exist to cause the problem.
- Demonstrate the problem is not simply an outlier.

Independent review/validation of each piece of data, assumption and conclusion is needed. It is useful to list all assumptions and conclusions and challenge each in turn. Re-checking the information, using different personnel is necessary.

Step 3.1B Cause Analysis using TRIZ Tools

Function Statement/Scientific Effect	
Step 1: Define the problem as a harmful SAO action (e.g. something corrodes metal).	
Step 2: Search scientific effects for the general function (how to corrode, corrosion, disintegrate, dissolve, etc.).	
Step 3: List all resources. (metal, air, air contaminants, water vapor, salt, temperature, static charge, etc.).	
Step 4: Create models of how the problem can be caused using the available resources.	

Nine Windows	
Step 1: Draw Nine Windows.	
Step 2: Write the problem in the center box and complete the diagram by entering the subsystem, system and supersystem components.	
Step 3: Consider various timescales and levels of system assess whether the problem is caused at that level/timeframe.	

Build a Functional Model	
Step 1: Determine what is to be modeled, the system the specific part of the system where the problem resides.	
Step 2: Build the model of how the system (or chosen parts) or process (or chosen steps) should operate.	
Step 3: Create ideas for how the problem can be caused.	

Create SFR List	

SFR List (Substance-Field Resource List)			
Internal System Resources	Substance	Field	Parameter
1 System Resources			
System components			
External Resources			
2 Supersystem Resources			
Environment where the problem is believed to reside			
Common to any Environment			
3 Surrounding Resources			
By products, Waste material			
Nearby systems			

RCS Statements	

Step 1: Build the Root Cause Statement (RCS) (this is a similar way we create an IFR statement in ARIZ). **RCS**: <X-resource> causes <the harmful affect	
Step 2: Substitute each raw resource into the X-resource part of the RCS statement.	

For example, we can use the example of a plasma etch tool used in the semiconductor industry to etch aluminum interconnecting metal. It comprises an electrode, chamber walls, process, gases, a robot transfer arm etc. An RCS list would look like this:

- The **electrode** causes the corrosion of the metal
- The **chamber walls** cause the corrosion of the metal

Etc.

Step 3: Create ideas based on the psychological inertia released by the RCS statement.	

By considering each RCS statement in turn, we release psychological inertia leading to many new ideas and potential solutions.

Step 4: If the above "raw" resources do not provide a solution, try the following secondary resources in the RCS statement: **RCS**: <X-secondary resource> causes <the harmful affect	

Secondary Resources:

Use a mixture of resources.

Use a mixture of resources	

Consider using empty space or a combination of a substance with empty space to solve the problem (Empty space can mean space, gap, cavities, bubbles etc).

Use empty space (gaps, bubbles, cavities, voids etc.).	

Consider using derivatives (or a mixture of derivatives with empty space)
A derivative is a derived resource, it is a changed resource that has changed phase, is a part of materials, burnt remains, decomposed materials, molecules etc anything we can derive from an existing resource or resources.

Use Derivatives	
Use derived resources with empty space (gaps, bubbles, cavities, voids etc.).	

Hidden resources These are resources that are present but often assumed to not be present. For example contamination level in water, foreign materials or constituents of a gas (instead of air, nitrogen, oxygen, argon etc, instead of wood - wood also contains moisture, etc).

Use hidden resources	
Use hidden resources (contaminants, foreign materials, constituents, etc.).	

Step 5: Create ideas based on the psychological inertia released by the RCS secondary resource statement.	

348

DTC Operator	

Step 1: Create an SFR list (see above)	
Step 2: Apply the extreme "what if" question questions to each SFR raw and secondary resource. Does this suggest a root cause?	

What if we exaggerate not only the dimensions, time and cost to very large and very small but consider the extremes of any substance, field or parameter on our SFR list. By considering these extremes is a cause suggested?

For example
- If the tool temperature of were very high, how could this cause the problem?
- If the tool temperature of were very low, how could this cause the problem?

Smart Little People	

Step 1: Draw the problem.	
Step 2: Create ideas of what the little people (the may be many groups) are doing to cause the problem.	
Step 3: Create plausible models of root cause based on what the Smart Little People are doing.	

Although we draw the SLP performing useful and harmful actions at stages through a contradiction from one state to another, we can use the idea of SLP to prompt ideas for how to cause the problem to try to identify root cause. How would Smart Little People cause the problem? Make a list.

Subversion Analysis SA-1	

Step 1: State the problem.	
Step 2: List all the resources, including the secondary hidden and derived resources (see above SFR list) including their storage, manufacture, operation, the environment, people, procedures, etc.	
Step 3: Given the all the resources, create models for how to cause the problem.	
Step 4: Evaluate whether any of the models are consistent with the data and experiment results. Then try to determine whether they are the root cause.	

Play devil's advocate. Given all the resources, their storage, manufacture, operation, the environment, people, procedures, etc., how can you cause the problem? It is useful to task a group of content experts to provide a prioritized list of ways of how to cause the problem give all possible available resources.

If Cause is understood

349

3.2 Cause Known

3.2.1 Propose Ideas

A	
B	
C	
etc	

3.2.2 Brainstorm

A	
B	
C	
etc	

3.2.1 CEC-2

A	
B	
C	
etc	

If a solution cannot be found or a breakthrough solution is needed, use TRIZ, see TRIZICS Roadmap, Step 2 - Select Problem Type.

Step 4 Ideas/Solutions

Solutions Bank		
Source		Priority

Step 5 Implementation Plan

	Action	Priority	Projected Completion Date	Owner	Status
1					
2					
3					
Etc					

Appendix 8.7 Contradiction Problem Solving Flowchart - Simplified ARIZ

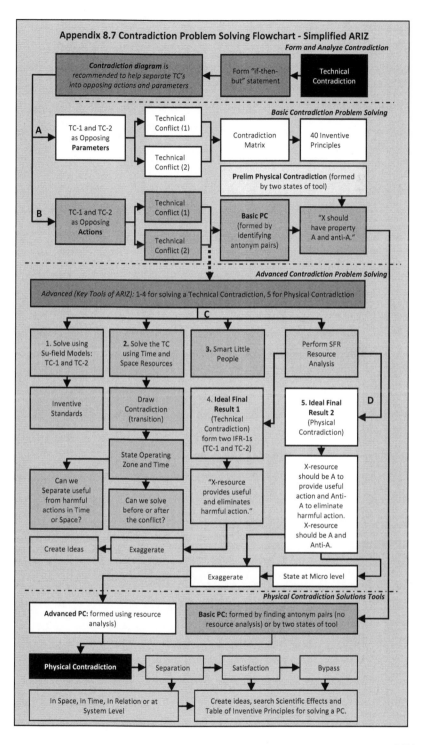

Appendix 8.8 Simplified ARIZ

We have combined the advanced methods above with the basic methods discussed in Chapter 9 Part 1 into a single algorithm for solving contradictions. It is a simplified form of ARIZ. It contains four main phases A (10), B (3), C (4) and D (1). The number of steps in each phase is shown in parenthesis. The algorithm describes the process for applying the **Contradiction Problem Solving Flowchart - Simplified ARIZ** shown in Chapter 9 Part 1 Section 2.5.

The "official ARIZ" version ARIZ-85C which many users difficult to follow is included in the Appendix 8.9. The main key problem solving tools of ARIZ are included in the Simplified ARIZ algorithm below including use of the Contradiction Matrix which is not used in ARIZ-85C.

Simplified ARIZ (18 Steps)

A. Form and Solve Technical Contradiction Algorithm using Basic Tools
B. Form and Solve the Physical Contradiction Algorithm using Basic Tools
C. Form and Solve the Technical Contradiction using Advanced Tools
D. Form and Solve the Physical Contradiction using Advanced Tools

A. Form and Solve Technical Contradiction Algorithm using Basic Tools

1. State the "if-then-but" contradiction statement and define the technical conflict statements for State 1 and State 2.
2. Identify the change between states TC-1 and TC-2.
3. Identify the tool (the thing, not a field) that delivers the change between states.
4. Identify the useful and harmful actions for TC-1 and TC-2 in terms of <State 1>Tool–Action–Object, and <State 2>Tool-Action-Object and input them into the diagram.
5. Identify the improving and worsening matrix parameters for both states.

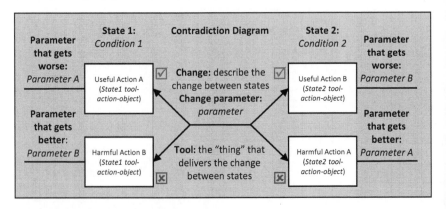

6. Sanity-check the diagram, TC-1 and TC-2 should be the mirror image of each other.
7. Make Graphical Representations of TC-1 and TC-2
8. For the Graphical Representations, ensure the interaction between the tool and object(s) is direct.
9. Solve the technical conflicts using inventive standards (Advanced Tool 1).
10 Solve the technical conflicts using the Contradiction Matrix and inventive principles

Note that solving technical contradictions using Su-field modeling and inventive standards is an advanced tool of ARIZ but we will use it in our algorithm for solving contradictions using basic tools.

B. Form and Solve the Physical Contradiction Algorithm using Basic Tools

11. For the chosen technical conflict, identify antonym pairs that support the action or parameters in opposition. We use a simple table.

For the chosen technical conflict:

A useful action or parameter: (to stop rain) Is a function of:	Anti-A harmful action or parameter : (to be difficult to carry) is a function of :	Parameter, Characteristic or Feature
big	small	size
wide	narrow	width

Identify items in column 1. By brainstorming what specific parameter, characteristic or feature that the improving action is a function of. In this case the improving action is to stop rain; this is a function of being big, being wide, being present.

Next, in column 2 (anti-A) we identify the antonyms of column 1. Now write the specific parameter, characteristic or feature the antonym pair describes. For example, big and small are measurements of size, size is input to column 3, wide and narrow are features of width, so input width to column 3. Continue this for all antonym pairs.

Now we input the information from the column into the Physical Contradiction statement below. If it makes sense then it forms a physical contradiction statement.

The <name the system or object> should be <A> to <state useful action or parameter> and <anti-A> to eliminate <state the harmful action or parameter>. The <name the system or object> should be <A> and <anti-A>.

For example,

The< umbrella> should be <**big**> to <stop rain> and <**small**> to eliminate <being difficult to carry>. The <umbrella> should be <**big**> and <**small**>.

Compile a list of physical contradictions and solve using the methods of satisfaction, separation and bypass (see below section 13).

12. Form the Preliminary Physical Contradiction Formation
When we identify the parameter of the change or what changes between the two states of a contradiction, we form a physical contradiction or the "tool." For example:

o   TC-1: **If** the bus is long **then** it carries many passengers **but** it can't turn corners.
o   TC-2: **If** the bus is short, **then** it can turn corners **but** it doesn't carry many people

The Preliminary Physical Contradiction is: The <bus> should be < long> and <short>.
The <tool> should be <State 1> and <State 2>
And solve using Step 3.

13. Solve the Physical Contradiction
When the Physical contradiction has been defined it is useful to use the template for solving a physical contradiction by separation, satisfaction and bypass.

353

Satisfaction	Solution
Can the contradictory requirements be solved by satisfaction?	

Try to determine a method to satisfy the problem - search scientific effects, use Table of Specific Inventive Principles to Solve Physical Contradictions.

Bypass	Solution
Can the contradictory requirements be solved by bypass?	

Try to determine a method to bypass the problem - search scientific effects, use Table of Specific Inventive Principles to Solve Physical Contradictions.

Separation in Space	Where
Where is there a requirement to be < A >?	
Where is there a requirement to be < anti-A >?	

If there is a requirement to be A and anti-A at the same location, then they cannot be separated in space.

Separation in Time	When
When is there a requirement to be < A >?	
When is there a requirement to be < anti-A >?	

If there is a requirement to be A and anti-A at the same time, then they cannot be separated in time.

Separation in relation	For what or whom?
For what is there a requirement to be < A >?	
For what is there a requirement to be < anti-A >?	

If there is a requirement to be A and anti-A for the same action or object, then they cannot be separated in relation.

Separation at system level (parts v whole)	Level
For what level of the system is there a requirement to be < A >?	
For what level of the system is there a requirement to be <anti- A >?	

If there is a requirement to be A and anti-A at all different system levels, then they cannot be separated at system level.

At this point, many of those who study and use TRIZ simply use the basic tools above to solve contradictions and never progress to the advanced tools for solving a contradiction. However, for more difficult inventive problems, the advanced tools (the tools of ARIZ) are useful.

C. Form and Solve the Technical Contradiction using Advanced Tools

14. Solving the Technical Contradiction using Time and Space Resources Algorithm (Advanced Tool 2)

Step 1 State the two TC's (technical conflicts).
Step 2 Complete the Contradiction Diagram.
Step 3 Draw the situation at T1, T2 and T3.
Step 4 Reference the drawing of T2, define: the Operating Time (OT) and Operating Space (OS).
Step 5 Re-state the two opposing requirements in T2.

354

Step 6 ask the questions
- Is it possible to separate the opposing requirements in space?
- Is it possible to separate the opposing requirements in time?
- Can the problem be solved by doing "something" in advance (during T-)?
- Can the problem be solved by doing "something" after (during T+)?

15. Algorithm for SLP (Advanced Tool 3)

Step 1: State the technical contradiction as two technical conflicts TC-1 and TC-2. Complete the Contradiction Diagram).
Step 2: Draw the current situation at T1, T2 and T3.
Step 3: From the drawings, identify the "should be." What should the little men be doing? (Assume you can add as many groups with special powers, etc) to solve the problem in TC-1 (T1), during the conflict (T2) or in TC-2 (T3). Step 4: Create several "should be" scenarios for each problem stage, if needed, to make as many ideas as possible
Identify practical ideas/solutions based on the various "should be" solutions.

16. Substance Field Resource Analysis

Complete an inventory of resources using the table below.

Substance-Field Resource List				
Internal System Resources	Substance	Field	Parameter	
1 System Resources				
Tool:				
Object 1:				
Object 2: (if present):				
External Resources				
2 Supersystem Resources				
Operating Space Environment				
Common to any Environment				
3 Surrounding Resources				
By-products, waste material				
Nearby systems				
Cheap or free available resources				

Secondary Resources

Resources 8.1 Use a mixture of resources.

Earlier we tried single resources in the SFR list. Now you can try combined (as many as you like).

Use a mixture of resources	

355

Resources 8.2 Consider using empty space or a combination of a substance with empty space to solve the problem (Empty space can mean space, gap, cavities, bubbles etc).

Use empty space (gaps, bubbles, cavities, voids etc.).	

Resources 8.3 Consider using derivatives (or a mixture of derivatives with empty space)

A derivative is a changed resource, change **phase**, parts of materials, burnt remains, decomposed materials, molecules etc anything we can derive from an existing resource or resources.

Use derived resources	
Use derived resources with empty space (gaps, bubbles, cavities, voids etc.).	

Resources 8.4 Hidden resources. These are resources that are present but often assumed to not be present or are not known to be present but actually are. For example: components of air, contamination level in water, foreign materials (wood also contains moisture) etc.

Use hidden resources	
Use contaminants, substances that may form be new mechanisms, etc.	

The SFR list is not a tool for solving a problem; it is an inventory of resources that will be used to solve the problem using Tools 4 and 5.

17. Algorithm for IFR-1 (Advanced Tool 4)

Step 1. State the technical contradiction.
Step 2. Complete the Contradiction Analysis and state TC-1 and TC-2 in terms of opposing actions. Identify the tool and object(s).
Step 3. Create a **Substance-Field Resource list** (see table).
Step 4. For TC-1 and TC-2 create the **IFR-1 statement**:

- TC-1 (state 2) "X-resource **itself** provides <useful action> and eliminates the <the negative action> without making the system more complex and without additional harmful consequences.
- TC-2 (state 2) "X-resource **itself** provides <useful action> and eliminates the <the negative action> without making the system more complex and without additional harmful consequences.

Step 5. Exaggerate the statements.
Step 6. Create the IFR-1 list by inserting each resource into the IFR-1 statement for both TC-1 and TC-2.
Step 7. List ideas.
Step 8. For more ideas use secondary resources:

   8.1 Combinations of resources.

356

8.2 Use emptiness (gaps, bubbles, cavities, voids, etc.).
8.3 Use derivatives (also combine derivatives with emptiness).
8.4 Hidden resources.

Step 9. Apply IFR-1 during T2. Consider if each resource can solve the problem during the transition between states (T2).

D. Form and Solve the Physical Contradiction using Advanced Tools

18 Algorithm for solving Physical Contradictions (IFR-2) – (Advanced Tool 5)

Step 1: Create the physical contradiction statement: X-resource (or part of it) should be "A" and "anti-A."

Where X-resource is a system, part of a system, or a component, "A" and "Anti- A" are an antonym pair (hot and cold, large and small, present and absent, etc.) of any property for X-resource.

- X-resource needs to have one property A to perform the useful action but the opposite anti-A to eliminate the harmful action.
- X-resource needs to have one property A but doesn't have that property.

Step 2: Create or refer to the Substance-Field Resource List used for IFR-1.
Step 3: Try to create one or more physical contradictions for each resource. Consider exaggerating the actions.
Step 4: Solve each physical contradiction by satisfaction, bypass or separation.

- Consider solutions during T1, T2 and T3. Even though we form only one physical contradiction (see above section in Chapter 9 Part 1. **Physical Contradiction - One Physical Conflict**, the problem may be solved in State 1, State 2 or during the transition between states).

Satisfaction	Solution
Can the contradictory requirements be solved by satisfaction?	

Try to determine a method to satisfy the problem - search scientific effects, use Table of Specific Inventive Principles to Solve Physical Contradictions.

Bypass	Solution
Can the contradictory requirements be solved by bypass?	

Try to determine a method to bypass the problem - search scientific effects, use Table of Specific Inventive Principles to Solve Physical Contradictions.

Separation in Space	Where
Where is there a requirement to be < A >?	
Where is there a requirement to be < anti-A >?	

If there is a requirement to be A and anti-A at the same location, then they cannot be separated in space.

Separation in Time	When
When is there a requirement to be < A >?	
When is there a requirement to be < anti-A >?	

If there is a requirement to be A and anti-A at the same time, then they cannot be separated in time.

Separation in relation	For what or whom?
For what is there a requirement to be < A >?	
For what is there a requirement to be < anti-A >?	

If there is a requirement to be A and anti-A for the same action or object, then they cannot be separated in relation.

Separation at system level (parts v whole)	Level
For what level of the system is there a requirement to be < A >?	
For what level of the system is there a requirement to be <anti- A >?	

If there is a requirement to be A and anti-A at all different system levels, then they cannot be separated at system level.

Step 5: Try to form each physical contradiction at the **micro level** (sometimes releases psychological inertia) and solve using satisfaction, bypass, separation.

Step 6: Try to form physical contradictions using the additional resources for combined resources, derived resources, emptiness hidden resources. And solve using satisfaction, bypass, or separation.

Step 7: Search for **scientific effects** use the Algorithm for How to Search for Scientific Effects to Solve a Physical Contradiction (see Chapter 7).

Appendix 8.9 ARIZ-85C

ARIZ-85C

Contents
1. *Introduction to ARIZ*
2. *ARIZ-85C Steps*
3. *Example Problem Description*
4. *ARIZ-85C Details*

Summary:
In this section we review and discuss ARIZ the Algorithm for the Solution of Inventive Problems.

1. Introduction to ARIZ

ARIZ is the Russian acronym for Algorithm Rezhenija Izobretatelskih Zadach known as the Algorithm for the Solution of Inventive Problems or ASIP in English. In TRIZ, an inventive problem is a contradiction. ARIZ is an algorithm for formulating and solving contradictions in order to create breakthrough solutions. ARIZ is part of TRIZ and was developed along with other TRIZ tools. The first version was created in 1968. The last "official" version approved by Altshuller is ARIZ-85C created in 1985. There are other more recent versions created by the followers of TRIZ, but ARIZ-85C is widely accepted as the "standard" version. When shown later versions, Altshuller commented the ARIZ-85C was "good enough." It is often quoted as being ARIZ-85B, this is because B is the third letter of the Cyrillic (Russian) alphabet, but in English it is ARIZ-85C. Sometimes it is also named as ARIZ-85V because B, the third letter of the Cyrillic alphabet is pronounced like a "v" as in vet in English.

ARIZ should not be used instead of the basic classical TRIZ solutions tools (searching scientific effects, Su-field modeling and inventive standards, technical contradiction and physical contradiction formulation and problem solving) which can be applied individually without ARIZ. "Standard problems," which Altshuller estimated make up about 85% of problems can be solved using the standard, classical TRIZ tools individually. ARIZ is designed to solve a "difficult" non-standard problem for which an inventive breakthrough solution is not achieved by standard TRIZ solutions tools.

In our TRIZICS Problem Solving Roadmap (Chapter 1 and Appendix 8.1), ARIZ is identified as a "solutions tool." It is not a problem definition tool, data gathering tool or tool for root cause detection of an engineering problem, the root cause and mechanism of a problem should be clearly understood before using ARIZ. It is not an analytical tool, other analytical tools like CEC, functional modeling etc that are appropriate to the problem type need to be applied first. It is a "solutions tool" for formulating and solving a problem into a contradiction and creating an inventive solution.

ARIZ is an algorithm made up of nine parts that combines several TRIZ tools for solving a contradiction into a sequential process. Each part has a number of steps and in total it contains more than 80 individual steps. It is designed to solve a contradiction by making minimal changes to the system - what is known as a "mini" solution. Mini in this case does not mean small, it means minimal. As we know from our study of ideality, solutions with minimal changes are more ideal since they introduce fewer resources (and incur less expense).

ARIZ starts with the formation of the problem as a technical contradiction and then re-formulates the problem into different formats in order to apply classical TRIZ tools and several other solutions tools to it. One important omission from ARIZ-85C is the 40 inventive principles used to solve a technical contradiction. In 1985 Altshuller abandoned solving technical contradictions using the Contradiction Matrix and 40 inventive principles. Instead he moved to creating Su-field models and applying the 76 inventive standards to solve a technical contradiction by addressing the harmful tool-action-object interaction of either technical conflict in a technical contradiction.

Twelve of the solution concepts listed in the 40 inventive principles are not captured by the inventive standards (numbers 2, 6, 7, 8, 9, 10, 12, 14, 17, 21, 27 and 33). These "missing principles" are useful in prompting ideas for solving contradictions. We therefore decided that it was important to keep the 40 inventive principles and Contradiction Matrix as part of the contradiction problem solving process. We recommend use of both the inventive standards (advanced tool) and 40 inventive principles (basic tool) for solving technical contradictions since they complement each other.

In Chapter 9 we discussed what we have classified as the basic and advanced tools for solving technical and physical contradictions. We created the Contradiction Problem Solving Flowchart – Simplified ARIZ (see Chapter 9 Part 1). The flowchart combines both the basic tools for solving contradictions omitted by ARIZ and the main tools used by ARIZ to solve a contradiction. The tools may be used individually; there is no requirement to perform them in sequence as is done in ARIZ-85C. Our recommendation is however to use the Contradiction Problem Solving Flowchart – Simplified ARIZ as a simple alternative to ARIZ-85C.

Using ARIZ can be very complex and confusing. Altshuller believed it needed a minimum of 80 hours of training. If the user prefers to use the actual ARIZ algorithm instead of the basic and advanced tools of the Contradiction Problem Solving Flowchart – Simplified ARIZ, then it is described below. It is recommended however to read Chapter 9 first which will make ARIZ easier to understand and follow.

ARIZ follows a highly structured step-by-step sequence containing nine parts. Each part has individual steps. Many expert users claim only Parts 1-4 are needed for problem solving and that 5-9 are "experimental" steps in development.

2. ARIZ-85C Steps

Part 1: Analyze the Problem
Part 2: Analyze the Problem Model (sometimes stated as analyze the resources)
Part 3: Define the Ideal Final Result and Physical Contradiction
Part 4: Mobilize and use the Substance-Field Resources
Part 5: Apply the Knowledge Base
Part 6: Change or Replace the Problem
Part 7: Review the Solution
Part 8: Apply the Solution
Part 9: Analyze the Problem Solving Process

The purpose of each part is described below.

To demonstrate how ARIZ is applied, we will use a simple example of a mousetrap for the user to reference as we proceed through the steps. Note that each example provided at each step is to illustrate what sort of solution should be created. It is not possible to provide a high level breakthrough solution for every tool for a single example. Typically we expect one or two **significant** breakthrough ideas when using ARIZ, not one for each tool. The user should appreciate the examples given below are to help the user understand the principle of what is required at each step. The response for a particular step that does not show a significant breakthrough in our example may provide a major breakthrough idea for the user's specific problem.

Note also that ARIZ is a multi step process that is basically a compilation of tools for creating breakthrough ideas. Each tool is expected to provide additional different solutions as you expose your problem to them. ARIZ does not identify a single solution then develop it as we proceed through the steps. Instead, we accumulate solutions. It may be sufficient to stop after Part 1 if you are satisfied with the solution obtained in Step 1.7. Or the user may decide to stop only after Part 2, and so on. Most problems have accumulated sufficient ideas by using the first four steps only. Any time a solution is generated and the user wishes to stop, we recommend that the user proceed to Part 7.

The major problem solving steps of ARIZ-85C are, Step 3.2, when we try to solve IFR-1 the technical contradiction, and Step 3.5, when we try to solve IFR-2 the physical contradiction. Typically problem solvers should focus their effort and time on searching for inventive solutions at these steps. In addition, the user should be aware that ARIZ is not to be done quickly. It may take days or even weeks to apply ARIZ to a single project, breakthrough ideas often emerge slowly as the process progresses.

Note that during the research of ARIZ-85C, several versions of it were found. No two versions use exactly the same steps or wording. Many of the steps suffer from poor and incomplete translation form the Russian. Our version is our best logical interpretation.

3. Example: Problem Description

A **mousetrap** is required to catch mice. The standard trap works well but it kills and injures the mice. How can the mousetrap be improved? Use ARIZ to find solutions.

Mousetrap Operation
In the loaded position the trapping arm is held by the holding bar (see drawings below). The holding bar is fixed at one end by the holding bar hinge. At the other end it is held by the cheese holder lever arm. If the mouse pushes down on the cheese, the lever arm is raised and the small frictional force keeping the holding bar in place is overcome by the force exerted on the trapping arm by the spring. The trapping arm flips from back to front capturing the mouse which is killed or injured. *The device need only capture the mouse by confining it such that it is trapped and cannot escape, the mouse does not need to be physically touched or damaged.*

Draw the system/problem

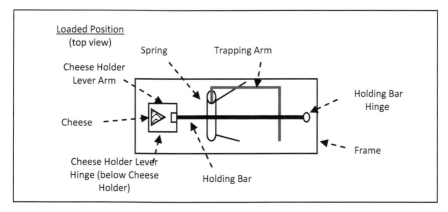

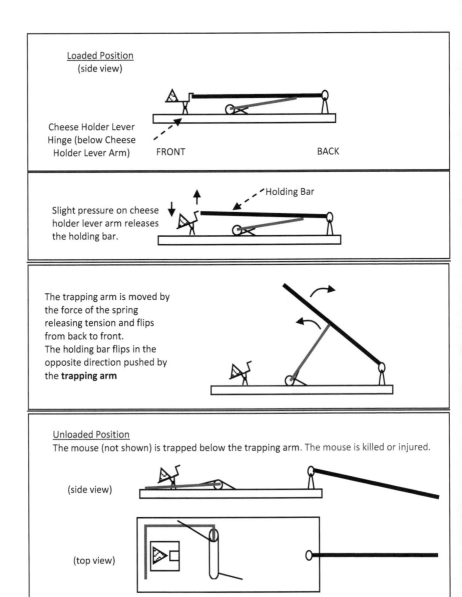

Loaded Position
(side view)

Cheese Holder Lever
Hinge (below Cheese
Holder Lever Arm)   FRONT   BACK

Slight pressure on cheese
holder lever arm releases
the holding bar.   Holding Bar

The trapping arm is moved by
the force of the spring
releasing tension and flips
from back to front.
The holding bar flips in the
opposite direction pushed by
the **trapping arm**

Unloaded Position
The mouse (not shown) is trapped below the trapping arm. The mouse is killed or injured.

(side view)

(top view)

4. ARIZ-85C

The first part of ARIZ-85C is to move from an initial vague problem to a well defined **problem model** of a technical contradiction to which we first apply the inventive standards.

362

1.1 Form the <u>mini problem.</u>
1.2 Define the conflicting elements (identify the tool and product(s)).
1.3 Describe the Graphical Representations of the technical conflicts.
1.4 Select the conflict that provides the best implementation of the main useful function.
1.5 Exaggerate the conflict.
1.6 Describe the <u>problem model.</u>
1.7 Apply the inventive standards.

1.1 Form the <u>mini problem</u>

The first challenge of ARIZ is to form the problem situation into a "**mini problem**."

To form a "mini problem" we first form our problem into a technical contradiction, we identify the two technical conflicts in State 1 and State 2. A mini problem is a way of stating a problem as **contradiction**. The useful actions of both technical conflicts are combined in a statement that includes the provision that the solution should require minimal changes to the system this statement is called the "mini problem"

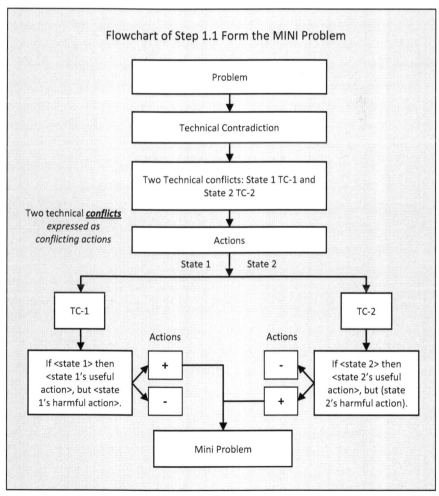

Flowchart of Step 1.1 Form the MINI Problem

363

The output of Step 1.1 should take the form:

TC-1: If there is < state 1> then, < the useful (+) action> but < the harmful (-) action>.

TC-2: If there is < state 2 > there is < the useful (+) action >, but < the harmful (-) action >.

Now form the "mini problem." It is necessary, with minimal changes to the system to <state the useful (+) action of TC-1> and <state the useful (+) action of TC-2>.

First, we state the positive and negative actions in State 1 and State 2, that is, we state technical conflict 1 (TC-1) and technical conflict 2 (TC-2). Then we state the mini problem which is a statement of the positive action of both states. The mini problem is the first statement of our objective, which to create both positive actions without compromise.

Let's complete ARIZ in template format. ARIZ doesn't normally have a template associated with it, but we will use one to keep organized.

Step 1.1 Form the mini problem	Example
There is a technical system for <state the **main useful function** > that includes <name main parts of system>.	There is a technical system for <trapping mice> that includes cheese, cheese holder lever arm, spring, trapping arm, holding bar hinge, frame, holding bar, cheese holder lever hinge, mouse, paint, air, person, room.
TC-1: If <state 1> then, <state the useful action> but < state the harmful action>.	If <the trapping arm force is high> then <the mouse is confined>, but <the mouse is damaged>.
TC-2: If < state 2 > then, < state the useful action >, but < state the harmful action >.	If <the trapping arm force is low> then <the mouse is not damaged>, but <the mouse is not confined>. The mouse is insufficiently conf8ined.
Form the "mini problem" - it is necessary, with minimal changes to the system to <state the **useful** action of TC-1> and <state the **useful** action of TC-2>.	It is necessary, with minimal changes to the system to <confine the mouse> and <not damage the mouse>.

1.2 Define the conflicting elements (identify the tool and objects(s)).	Example
Tool <name the tool> Object 1:<name the product> Object 2: <name the second product if present>	Tool: **trapping arm** (high and low force) Object: **mouse**

The tool is the object that delivers the change between the two states that directly interacts with the product(s). **The object** or objects are directly acted upon by the tool, it (they) are changed by the tool transitioning between states. There are either one or two objects. Identifying the tool and objects can be difficult without the assistance of the "**Contradiction Diagram**" discussed in Chapter 9. Although not part of ARIZ, we recommend the user creates a Contradiction Diagram to help define the tool and object(s), the states and the useful and harmful actions in each state. Also we recommend the user validates the interactions between tool and product(s) are direct. See also Chapter 9.

In our example of the mousetrap the change between states is force, we move from high force to low force and vice versa as we transition between states. The object is the mouse; it is what is changed by the tool. We could consider the change between states to be delivered by the spring; we could have a high tension spring and a low tension spring which changes the force on the

364

trapping arm. But the spring is not the tool, it does not directly interact with the object (mouse), it is the arm itself that directly interacts with the mouse. The arm is therefore the tool.

1.3 Describe the Graphical Representations of the technical conflicts	Example
TC-1 Graphical Representation	TC-1 confines / Tool: High Force Arm damages Mouse
TC-2 Graphical Representation	TC-2 confines / Tool: Low Force Arm does not damage Mouse

Describing the state of the tool is of key importance when identifying the tool, object(s) and useful and harmful actions. Using the state of the tool as an adjective before naming the tool helps clarify the two contradictions, without the "adjective" it can become confusing. In our example we describe the tool as the "high force" arm and the "low force" arm.

Note also that the useful action for TC-2 is a null positive action "the low force arm **does not** damage the mouse."

1.4 Select the conflict that provides the best implementation of the main useful function.	Example
Select TC-1 or TC-2	TC-1 The main useful function of the mousetrap is to confine mice, which is best delivered by TC-1 when there is a high force arm.

At this point ARIZ directs us to choose a conflict TC-1 or TC-2. The recommendation is to first choose the technical conflict that best delivers the **main useful function** of the system this often leads to solutions that improve the existing system, selecting the other usually leads to more radical changes or a new **alternative** design. In our mousetrap example, to try to stop the damage to the mouse when the spring is strong (existing situation) would involve less radical change than to find a way of confining the mouse when the spring is weak. Note that it is not necessary to choose a conflict since it is not clear which solution might be best, but working in parallel can be complex ,so we recommend that the user chooses only one initially. In ARIZ-85C if a solution is not found using our first choice, we are instructed to choose the other technical conflict in Step 6.3.

1.5. Exaggerate the conflict.	Example
For the chosen TC (TC-1 or TC-2). State the **un-exaggerated conflict** If <describe state 1> then <describe the positive action>, but <describe the negative action>.	**un-exaggerated conflict** TC-1: If <the arm has high force> then <the mouse is confined> but <the mouse is damaged>.
For the chosen conflict state the **exaggerated conflict** If <describe state 1> then <describe the positive action>, but <describe the negative action>.	**exaggerated conflict** TC-1: If <the arm slams down with tremendous force> then <the mouse is completely confined> but <the mouse is severely damaged>.

By exaggerating, (aggravate, strengthen) the conflict we release psychological inertia. For all subsequent processing of the selected conflict using ARIZ we will use the exaggerated version of the conflict.

1.6 Describe the problem model	Example
State the exaggerated conflict (from 1.5) If <describe state 1> then <describe the positive action>, but <describe the negative action>.	TC-1: If <the arm slams down with tremendous force> then <the mouse is completely confined> but <the mouse is severely damaged>.
State the conflicting pair (the tool and the object(s)) of the exaggerated conflict. Include the description of the exaggerated state of the tool Tool: Object 1: Object 2: Describe the conflicting pair in a simple sentence	There are: Tool: an arm that has tremendous force Object 1: a mouse **There are: an arm with tremendous force and a mouse.**
State the **Problem Model in the form:** It is necessary to find an X-element that provides or maintains <the exaggerated useful action> and completely eliminates <the negative action>.	**Problem Model** It is necessary to find an X-element that <provides complete confinement of the mouse> and <eliminates all damage to the mouse>.

The problem model is a concise statement of the problem in the chosen technical conflict. It introduces the idea of "something" an x-element that solves the problem. The problem model should state what X-element needs to do to solve the problem. In this case, confine the mouse without damage.

Now we are going to put all the above aside and come back to it later in Part 3 after we have compiled a list of resources in Part 2.3 Now let's try to solve the technical contradiction by addressing the harmful action of the chosen technical conflict. Altshuller chose to solve technical contradictions using inventive standards rather than the Contradiction Matrix and 40 inventive principles.

366

1.7 Apply the inventive standards	Example
Try applying the inventive standards to the problem model. For the exaggerated chosen technical conflict, try to find ways to improve the harmful, insufficient, and excessive or absent action from the Graphical Representation.	TC-1 Completely confines Tool: Severely High Force damages Mouse Arm The problem is to address the **harmful interaction:** high force arm – severely damages – mouse.

Although not part of ARIZ, we recommend using the Inventive Standards Flowchart – see Chapter 8 and Appendix 5. In this example of the mousetrap, we have a harmful interaction and we are directed to Class 1.2, 3 and 5.

Ideas Solutions
Inventive Standard 1.2.1 Introduce a new foreign substance S3 between S2 and S1. Idea, place a sponge on the arm so the mouse is not damaged but is held securely.
Inventive Standard 1.2.2 Add a new modified substance between S2 and S1. Solutions include add in "nothing" holes, gaps, etc. Idea: move the trapping area to beneath the surface of the frame, the mouse will not be damaged but will be held.
There are many more ideas that can be prompted by the inventive standards. Try creating some using groups 1.2, 3 and 5 of the inventive standards.

If satisfied with the solution, move to Part 7. If at any time, in any step of ARIZ, a satisfactory solution is found, proceed to Part 7- Review the Solution.

Part 2: ANALYZE THE PROBLEM MODEL

Sometimes titled Analyze the Resources, Part 2 is mainly for making an inventory of available resources (space, time, substances and fields) that can solve the problem. In Part 2 we also try to solve the problem by separation of the opposing useful and harmful actions of the chosen technical conflict in space and time, or trying to prepare a solution in advance or recover the problem after the conflict.

2.1 Define the operating space (OS)
2.2 Define the operating time (OT)
2.3 Define the substance and field resources (SFR)

2.1 Define the operating space (operating zone)	Example
The operating space is the region of space where the conflict occurs.	For TC-1 the operating space is where the mouse is confined and damaged.

It is important to know the operating space is also known as the operating zone and that the abbreviations OS (or OZ) and OT and SFR mentioned above, are frequently used in TRIZ and ARIZ and are important to the users knowledge.

It is useful to draw the operating space and identify it.

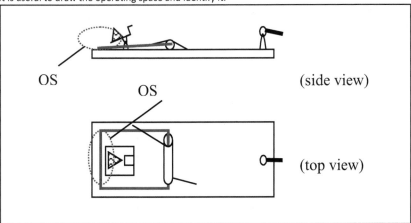

OS   (side view)

OS   (top view)

Is it possible to separate the useful and harmful actions in space?	For TC-1 The useful action is arm confined mouse. The harmful action is arm damages mouse. Can they be separated in space? Yes, to capture the mouse the arm must merely confine the mouse, it does not need to touch (damage) it.
If useful and harmful actions can be separated in space, create ideas/solutions,	Idea (using Principle 9 prior counter-action) recess the trapping area in a hole so the mouse is not damaged.

Although not part of ARIZ it is useful to think of separating useful and harmful actions in the same way we used to create ideas for separating conflicting physical requirements of a physical contradiction in space and time. The table "Specific Inventive Principles to Solve Physical Contradictions." is shown below.   See Chapter 9 Part 1 or Appendix 7, it shows inventive principles useful for separating in space and separating in time.

2.2 Define the Operating Time (OT)	Example
The operating time is the time when the conflict occurs.	For TC-1 the operating time is when the mouse is captured by the high force arm and when the mouse is damaged.
Is it possible to separate the useful from the harmful action in time?	In this case it is not possible to separate when the arm captures the mouse from when the mouse is damaged.
If useful and harmful actions can be separated in time, create ideas/solutions,	N/A
Is it possible to prepare a solution to the problem in advance during T2?	Mouse could try wearing a crash helmet.
Is it possible to provide a solution after the conflict, during T3 something that can nullify or recover from the harmful effect?	In this case no, the damage to the mouse is not repairable.

ARIZ-85C defines three times, the time before (T2) during (T1) and after (T3) the conflict.

Time

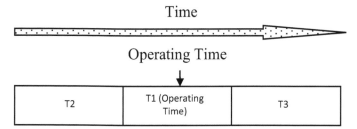

Operating Time

T2	T1 (Operating Time)	T3

The convention in ARIZ-85C is to use the sequence T2 – T1 – T3, not T1 –T2 –T3 as time proceeds. T1 is the operating time, the time of the conflict, T2 is the time before the conflict, and T3 the time after the conflict.

Note that these do not match our definitions in Chapter 9. In Chapter 9 we defined T-, T1, T2, T3 and T+. This is because the conflict really occurs at and through the transition between TC-1 and TC-2 and in our analysis in Chapter 9 we drew the problem situation at TC-1, at TC-2 and during the transition (the time of maximum conflict) instead of trying to produce a single drawing.

2.3 Create a substance-field resource list (SFR)	Example
Define the substance and field resources of the analyzed system, the external environment and its products	See below.

Using the table below, we make an "inventory" of all the substance and field resources, including their parameters. Note that substance has the same meaning as in Su-field modeling it is any physical object, any "thing" of substance. The same list applies for both TC-1 and TC-2; the same resources should be available for both.

- A substance is any object or material present that is not a "field."
- A field is as we define it in Su-field modeling includes all the strict definitions used for fields in science (weight, pressure, temperature, magnetic properties, etc.). But we also include a wider type of definition: examples are taste, pressure field, smell (olfactory) field, stickiness field, biological, chemical, acoustical, optical field, etc.
- A parameter is a property or feature or characteristic of a substance or field can be quantified or measured.

Note that the substance-field resource list is a compilation of "raw" single resources. Later, if needed, we will add to our list of resources by considering combinations of these resources, resources that can be derived from them, etc.

Substance and Field Resource List (SFR)			
Internal System Resources	Substance (thing)	Field	Parameter
1 System Resources			
Tool: trapping arm	trapping arm	force of arm	Shape of arm Flexibility of arm Material composition of arm The direction of motion of the trapping arm
Product 1: mouse	mouse	thermal	Size, color, intelligence

369

		(temperature)	of mouse.
Product 2: (if present): N/A	(Not Present)		
External Resources			
2 Supersystem Resources	Cheese Cheese holder Lever arm Lever arm hinge Frame Spring Holding bar Holding bar hinge Floor the trap is on Room the trap is in	Thermal (temperature) Mechanical force (tension) of spring. Smell of Cheese (olfactory field) Gravity Earth's Magnetic Field	Color, shape, size, weight, porosity, hardness, type, internal structure, composition temperature, flexibility, dynamicity, uniformity, position, direction of motion, etc.
Operating Space Environment	Local temperature	Ambient light	Color
Common to any Environment	Air Contaminants of air	Gravity Earth's magnetic field	Atmospheric pressure Ambient Temperature Ambient light
3 Surrounding Resources			
By products, waste material	Dead mice. Used traps.		
Cheap or free available resources	Nearby furniture Water supply is nearby Cardboard tube Old fishing line		

In this section we formulate IFR-1. Ideal Final Result 1 is a formulation of the technical conflict. In Part 1 we created a problem model, then left it aside while we compiled an inventory of resources in Part 2. We now return to the problem model that we left aside at the end of Part 1.

The problem model took the form of an X-element statement:

"It is necessary to find an X-element that provides/maintains <the exaggerated useful action> and eliminates <the exaggerated negative action>."

In Part 2 we created an inventory of "raw" resources. In Part 3 we combine the problem model with the resource list to form a list of ideal solutions statements. This list is called IFR-1. These are statements that prompt ideas to solve the chosen technical conflict (and hence are statements to solve the technical contradiction).

IFR-1 is the technical contradiction formulation of the problem.

3.1 Formulate the IFR-1

3.2 Formulate the Intensified IFR-1

3.3 Form the Macro Level Physical Contradiction

3.4 Create Micro Level Physical Contradiction

3.5 State IFR-2

3.6 Apply Inventive Standards to the Physical Contradictions identified in the IFR-2.

3.1 Formulate IFR-1	Example
Formulate the wording of the ideal final result (IFR-1): X-element *itself* without complicating the system and without causing harmful side effects, eliminates <indicate the exaggerated harmful action> during the operating time in the operating space while preserving the ability to provide <indicate the exaggerated useful action>.	IFR-1: X-element *itself*, without complicating the system and without causing harmful side effects, eliminates <all damage to the mouse> <during capture> <where the mouse is confined> while preserving the ability to <completely confine the mouse>.

The Ideal Final Result simply put is, I changed nothing in my system and the problem is solved.

The next step is to remove the X-element and replace it with each item on the SFR list. This is a very powerful step for releasing psychological inertia. The intensified IFR simply refers to inputting resources from the SFR list in place of the phrase "X-component" to create the IFR-1 list. This is called intensification

3.2 Formulate the Intensified IFR-1	Example
Try each item on the SFR-list in place of the "x-element"	**The trapping arm itself**, without complicating the system and not causing harmful side effects, eliminates <all damage to the mouse> <during trapping> <where the mouse is captured> while preserving the ability to <completely capture the mouse>.
	Force of the trapping arm itself, without ....etc.
	Flexibility of the trapping arm itself, without ....etc.
	Shape of the trapping arm itself, without ....etc.
	Material composition of the trapping arm itself, without ....etc.
	The direction of motion of the trapping arm itself, without ....etc.
	The mouse itself, without ....etc.
	The temperature of the mouse itself, without ....etc.
	Size of the mouse color itself, without ....etc.
	The color of the mouse itself, without ....etc.
	Intelligence of the mouse itself, without ....etc.
	The cheese itself, without ....etc.
	Etc.

371

It is not necessary to write down each intensified IFR statement but each item on the list should be individually considered, this often leads to many new ideas.

Creating ideas and solutions prompted by 3.2 is a major step. The user should expend much thought and effort to solve the problem now and at Step 3.5. We also recommend the user considers combinations of resources; derivatives; etc. at this stage and not wait until Step 4.3 to 4.7 when ARIZ-85C does consider the combinations, etc.

Ideas/Solutions Prompted by 3.2
The flexibility of the trapping arm itself, without complicating the system and not causing harmful side effects, eliminates <all damage to the mouse> <during capture> <where the mouse is confined> while preserving the ability to <completely confine the mouse>. **Idea is to make the bar at the end of the trapping arm of elastic rubber. It will stretch and not damage the mouse.**
The shape of the arm itself, without complicating the system and not causing harmful side effects, eliminates <all damage to the mouse> <during capture> <where the mouse is confined> while preserving the ability to <completely confine the mouse>. **Make the arm into the shape of a bell to confine but not touch the mouse** 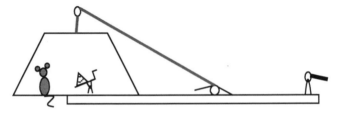
The material of the arm itself, without complicating the system and not causing harmful side effects, eliminates <all damage to the mouse> <during capture> <where the mouse is confined> while preserving the ability to <completely confine the mouse>. The arm is made of a large sponge.
The direction of motion of the trapping arm itself, without complicating the system and not causing harmful side effects, eliminates <all damage to the mouse> <during capture> <where the mouse is confine> while preserving the ability to <completely confine the mouse>. The trapping arm should miss the mouse but it should surround the mouse and hold it. It could rotate around the mouse rather than strike it.
The mouse itself, without complicating the system and not causing harmful side effects, eliminates <all damage to the mouse> <during capture> <where the mouse is confined> while preserving the ability to <completely confine the mouse>. The mouse could trap itself. We could make a box with a door that opens only one way. The mouse attracted by the cheese would walk in and trap itself.
The cheese itself, without complicating the system and not causing harmful side effects, eliminates <all damage to the mouse> <during capture> <where the mouse is confined> while preserving the ability to <completely confine the mouse>. Make the cheese sticky to hold the mouse and so we can remove the arm.
Gravity itself, without complicating the system and not causing harmful side effects, eliminates <all damage to the mouse> <during capture> <where the mouse is confined> while preserving the ability to <completely confine the mouse>. The mouse "walks the plank" and tips itself into a bucket by gravity. The mouse reaches the plank via a path made by the user.

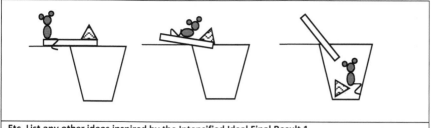

Etc. List any other ideas inspired by the Intensified Ideal Final Result 1

If solution is satisfactory, move to Part 7.

Up until now from Parts 1 through to the end of Step 3.2 we have formed and tried to solve our problem as a **technical contradiction**. Now, for the rest of Part 3 we try to formulate and solve our problem as a **physical contradiction**.

3.3 Form the Macro Level Physical Contradiction	Example
Formulate the wording of the physical contradiction at the macro level: X-element (or part of X-element) in the operating space during the operating time has to <indicate one physical requirement e.g. be soft> in order to perform <indicate one of the conflicting actions> and should <indicate the opposite physical macro state e.g. be hard> to perform <indicate the other conflicting requirement or demand>.	Macro Level Physical Contradiction: **X-element (or part of it)**, without complicating the system and not causing harmful side effects, <during capture> <where the mouse is confined> has to be A to eliminate <all damage to the mouse> and anti-A to <completely capture the mouse>.

We must now consider each substance and field resource (or part of each resource) from out list and try brainstorm antonym pairs for each to define physical contradictions. More than one antonym pair can be created for a resource.

X-element (resource from SFR list)	Antonym (A) <to eliminate damage to the mouse>	Antonym (Anti-A) <to completely confine the mouse>
The trapping arm	Absent	Present
Force of the trapping arm	Low	High
Flexibility of the trapping arm	High	Low
Shape of the trapping arm	-	-
Material composition of the trapping arm	Soft	Hard
The direction of motion of the trapping arm	-	-
The mouse	Absent	Present
The temperature of the mouse	-	-
Size of the mouse	-	-
The color of the mouse	-	-
Intelligence of the mouse	-	-
The cheese	-	-
Etc.		

We proceed through the list identifying more physical contradictions. Usually there are several macro physical contradictions formed that we could solve.

Macro Physical Contradiction
The operating space <where the mouse is confined> should provide **<a present and absent trapping arm>** by itself during the operating time <during capturing>.
The operating space <where the mouse is confined > should provide **<a hard and soft trapping arm>** by itself during the operating time <during capturing >.
The operating space <where the mouse is confined > should provide **<a trapping arm composed of soft and hard material>** by itself during the operating time <during capturing >.
The operating space <where the mouse is confined > should provide **<a present and absent mouse>** by itself during the operating time <during capturing >.
Etc.

Let's take the first one. The trapping arm must be present and absent.

Normally we wait until after IFR-2 is formed in 3.5 to find solutions, but let's do an example now.

We can solve by separation, satisfaction and bypass. Although not part of ARIZ, it is useful to refer to the table; Specific Inventive Principles to Solve Physical Contradictions (see Step 2.1).

It is useful to describe the transition as the conflict as it proceeds in time:
As the arm comes down with high force it must be absent when it contracts the mouse and when it confines the mouse it must be present.
Solution

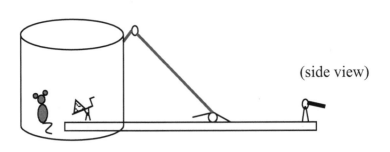

(side view)

Use a hollow cylinder shaped arm. As it falls, the arm is absent relative to damaging the mouse and present relative to confining the mouse. We have separated in space.

For every macro physical contradiction we can try to form a micro physical contradiction. We try to state each macro contradiction at the micro level (level of particles, grains, molecules, ions, atoms, electrons, etc.). Altshuller believed significant breakthrough ideas can be achieved by solving the physical contradiction at the micro level. Note that it is not always possible to state the problem at the micro level.

Let's take the second Macro Level Physical Contradiction from the list above as an example:
The operating space <**where the mouse is confined** > should provide **<a hard and soft trapping arm>** by itself during the operating time <**during capturing** >.

374

3.4 Form the Micro Level Physical Contradiction	Example
Formulate the wording of the Physical Contradiction at the Micro Level X-element (or part of X-element) in the operating space during the operating time has to <indicate one physical condition or action of particles of matter, ions, molecules etc.> to provide <one macro state from Step 3.3 e.g. to be soft> (to eliminate the harmful action) and <indicate the opposite physical condition or action of particles of matter, ions, molecules etc.> to provide <the opposite macro state from Step 3.3 e.g. to be hard>(to provide or maintain the useful action)	Using the second Macro Level Physical Contradiction from the list above: **The trapping arm (or part of it)**, without complicating the system and not causing harmful side effects, <during capture> <where the mouse is confined> should be **made of free flowing particles>** to be **<soft>** to eliminate **<all damage to the mouse>** and should be **<made of tightly bound particles.>** to be <solid> to <completely capture the mouse>.

The objective is to state each macro level physical contradiction at the micro level, in physical condition or action of particles of matter, ions, molecules, etc.

We normally wait until the IFR-2 has been stated to solve the micro physical contradiction. But let's complete one example.

We can solve by separation, satisfaction and bypass. Although not part of ARIZ, it is useful to refer to the table; Specific Inventive Principles to Solve Physical Contradictions (see Step 2.1).

Solution

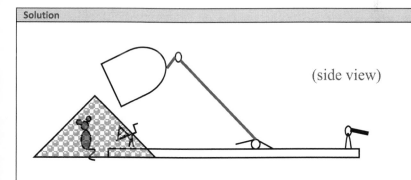

(side view)

The "trapping arm" is made of a large volume of grains that confine but do not harm the mouse. The arm dumps the grains onto the mouse using a cup.

IFR-2 is a compilation of all of the macro and micro level physical contradiction statements listed by substance-field resource (just as IFR-1 was a compilation of the technical contradictions) with different wording.

3.5 State IFR-2	Example
Formulate the wording of the IFR-2 at the Macro Level: The operating space <state> should provide <indicate the opposing macro or micro physical states> itself during the operating time <state>.	Macro Level Physical Contradiction: The operating space <where the mouse is captured> should provide **<indicate the opposing macro or micro physical states>** by itself during the operating time <during capturing>.

375

In the table below, we will list the IFR-2's from the macro physical contradiction list then the micro physical contradiction list. Note that ideally only one IFR-2 is selected from the combined lists below. The rest of ARIZ is written as if only one IFR-2 has been selected. Although there is actually no way to know which resource of part of it might form the physical contradiction that best solved the problem. Note also that the lists are "separate" in the sense that row 1 of the macro list does not correlate to row 1 of the micro list.

IFR-2

IRF-2's (macro list)
The operating space <where the mouse is captured> should provide **<a present and absent trapping arm>** by itself during the operating time <during capturing>.
The operating space <where the mouse is captured> should provide **<a high and low flexibility trapping arm>** by itself during the operating time <during capturing >.
The operating space <where the mouse is captured> should provide **<a trapping arm composed of soft and hard material>** by itself during the operating time <during capturing >.
The operating space <where the mouse is captured> should provide **<a present and absent mouse>** by itself during the operating time <during capturing >.
etc.

IRF-2's (micro list)
The operating space <where the mouse is captured> should provide **<a trapping arm with fluid free flowing particles and solid tightly bound particles >** by itself during the operating time <during capturing >.
The operating space <where the mouse is captured > should provide **<a trapping arm with gas-like free moving particles and solid tightly bound particles>** by itself during the operating time <during capturing >.
The operating space <where the mouse is captured > should provide **<trapping arm material composed of powder loose particles and solid tightly bound particles>** by itself during the operating time <during capturing >.
The operating space <where the mouse is captured > should provide **<a mouse with gas-like free moving particles** and solid **tightly bound particles>** by itself during the operating time <during capturing >.
etc.

Ideas and concepts of solutions for solving the physical contradictions listed in IFR-2 can now be created.

Creating ideas and solutions prompted by 3.3 to 3.5 is a major step. The user should expend much thought and effort to solve the problem now. Some of the formulations might not be viable to pursue, for example, creating a vapor mouse, but the exercise is still valuable, it may prompt a radical new idea.

Note that the way ARIZ-85C is written, it's as if there is only one IFR-2 for the subsequent steps. In fact we have a number of IFR-2's to work with because the IFR-2's are derived from the macro and micro physical contradictions created above in Steps 3.3 and 3.4. For a real example all should be pursued through the subsequent steps of ARIZ because it is not known which resource or part of it might form the physical contradiction that best solves the problem. In subsequent steps relating to IFR-2 we go back and forth selecting different IRF-2's to help illustrate the example solution.

376

The next step helps us to find solutions to the physical contradictions stated in IFR-2 by using inventive standards. In this example for illustrative purposes we will choose the third Macro IFR-2.

The operating space <where the mouse is captured> should provide **<a trapping arm composed of soft and hard material>** by itself during the operating time <during capturing >.

3.6 Apply Inventive Standards to the Physical Contradictions identified in the IFR-2	Example
State the IFR-2 The operating space <state> should provide <indicate the opposing macro or micro physical states> itself during the operating time <state>. Apply inventive standards.	The operating space <where the mouse is captured> should provide **<a trapping arm composed of soft and hard material>** by itself during the operating time <during capturing >. Inventive Standard 1.2.1. There is a useful and harmful interaction between S2 (the trapping arm) and S1 (the mouse). F is a mechanical field. Solution Add S3 between S2 and S1
List Ideas	The idea is to place a sponge under the arm. It will prevent damage to the mouse and confine the mouse.

Note this is not the same problem as 1.7 for which we used inventive standards to improve the harmful action of the chosen technical conflict.

Part 4: MOBILIZE AND USE SUBSTANCE-FIELD RESOURCES

Part 4 includes "Smart Little People" and "one step back" tools for releasing psychological inertia to prompt ideas. It also deals with how to make more of the resources, consider combinations or derivatives of resources instead of just the raw resources, consider using "nothing" voids, etc.

4.1 Modeling with Smart Little People.
4.2 One Step Back
4.3 Combined Resources
4.4 Using Empty Space
4.5 Using Derivatives
4.6 Using Electric Fields
4.7 Using a Field and Substance additive for which the substance is responsive to the field.

Modeling with Smart Little People was discussed in Chapter 9. Note that ARIZ requires drawing the Smart Little People only for the chosen technical conflict. This is not possible if the technical conflict contains a null positive action. It is recommended the user draw the problem as we did in Chapter 9, during TC-1, during the transition and during TC-2.

4.1 Modeling with Smart Little People	Example
Algorithm for SLP 1. State the technical contradiction as two technical conflicts TC-1 and TC-2 defined in Part 1. 2. Draw the current situation for the chosen conflict states TC-1 or TC-2 (we recommend drawing both states and during the time of transition which is not part of ARIZ). 3. From the drawings, identify the "should be." What should the little men be doing? Assume you can add as many groups with special powers, etc. to solve the problem in TC-1, during the conflict or in TC-2. Create several "should be" scenarios for each problem stage, if needed, to make as many ideas as possible. 4. Identify practical ideas/solutions based on the various "should be" solutions.	It is expected that the user would complete these drawings. See Chapter 9 for examples.

Published versions of Step 4.2 are conflicting, some refer to this step as one step back from the IFR-2, some from IFR, and some simply call it one step back. The main concept is to find a solution by "almost solving" the problem or "stepping back" from an ideal solution then solving the residual problem. It does not mean metaphorically taking a step back to see the "big picture."

4.2 One Step Back	Example
If there is an idea for how to solve a problem but it is not known how to achieve it. Try taking one step back from the IFR (the solution), nearly solve the problem. Then work on solving the residual often simpler problem. 1. State the target solution (an idea from IFR-1, IFR-2, SLP, or any other idea that could solve the problem but can't be achieved or is difficult to implement). 2. Define a slightly deteriorated solution. 3. Identify a solution to the residual problem	1. There is water gushing from dam wall, you cannot get close enough to stick on a patch to stop it because of the force of the water. The target solution is to stick the patch on the wall over the hole. 2. Attach a wide circumference hose around the gushing water allowing the water to flow freely through the hose resulting in no force. The hose does not stop the water but it does direct it. 3. Solve the simpler problem of how to shut off the hose.

Note that there is a second common description for "one step back." It is to deteriorate the problem situation then to solve the worsened situation. For example instead of one hole, the problem solver could make lots of holes. The dam would drain more quickly reducing the force of the water, it is easier to place patches on these holes than the original one. It is recommended both definitions of "one step back" be considered. It is not important what is correctly defined in ARIZ, it is important to create ideas no matter what prompted the idea.

4.3 Combined Resources	Example
In Part 3 we considered only "raw" resources from the SFR list. Consider Part 3 again using combinations of resources from the SFR list, not just one at a time.	For IFR-1: The **direction and shape** of the trapping arm without complicating the system and not causing harmful side effects, eliminates <all damage to the mouse> <during capture> <where the mouse is confined> while preserving the ability to <completely confine

the mouse>.

The idea is to create a cylindrical trap with rotating wall as the "arm." The mouse is lured through the door, it springs the trap, the arm is shaped as an outer wall that rotates, thus trapping the mouse inside

Loaded trap

Sprung trap

The mouse can be disposed of without damage.

4.4 Using Empty Space	Example
Consider empty space or a combination of a substance with empty space to solve the problem (empty space can mean space, gap, cavities, bubbles, voids, etc).	Instead of a trapping arm, use a trapping box or net which has a safe cavity to trap the mouse.

Consider using empty space or a combination of a substance with empty space to solve the problem. For example, bend the trapping arm to create a space to trap the mouse.

4.5 Using Derivatives	Example
In Part 3 we considered only "raw" resources from the SFR list Consider using derivatives (or a mixture of derivatives with empty space).	Attach a dead mouse from a previous trapping to the trapping arm to soften the blow to the live mouse.

A derivative is a changed resource, change phase, parts of materials, burnt remains, decomposed materials, molecules, etc. anything we can derive from an existing resource or resources.

4.6 Using Electric Fields	Example
Consider using an electric field or two interacting electric fields instead of a substance. This means instead of using an object or part of an object, try using electric field. Of course you have to introduce a way of creating the field.	Instead of the trapping arm, provide an electric shock to "stun" the mouse via the frame (base).

4.7 Using a Field and Substance additive for which the substance is responsive to the field.	Example
Consider using a field and substance additive that is responsive to the field. Use things like, thermal field with shape memory metal, ultraviolet light with phosphor, magnetic field with ferro-magnetic material and so on.	Feed the mice nearby material that makes them glow in ultraviolet light. Later, catch them with a net by using an ultraviolet flashlight to search for them in the dark. Of course, even if you can see the mouse it can be difficult to catch.

379

In many cases the problem is solved by Part 4 of ARIZ. If no satisfactory solution has been achieved by 4.7, then move to Part 5. If the problem has a solution by Step 4.7 then miss out Part 5 and 6 and move on to Part 7 "Review the Solution."

The main purpose of Part 5 is to use existing knowledge.

5.1 Apply Inventive Standards
5.2 Apply Problem Analogues
5.3 Apply Principles for Resolving Physical Contradictions
5.4 Apply the directory (database) of Physical Effects.

The purpose of Part 5 is to apply all accumulated information in the TRIZ knowledge base

5.1 Apply Inventive Standards	Example
Consider solving the problem using inventive standards just as we did with 3.6, using the wording of the problem of IFR-2 but this time take into account the SFR resources (derived, combined etc) that were identified in Part 4.	The chosen IFR-2 in 3.6 was: The operating space <where the mouse is captured> should provide <a trapping arm composed of soft and hard material> by itself during the operating time <during capturing >. Inventive Standard 1.2.1 There is a useful and harmful interaction between S2 (the trapping arm) and S1 (the mouse). Solution Add S3 between S2 and S1 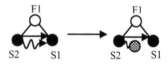 Instead of adding a sponge under the arm as we did in 3.6. We can add a bunch of fishing line which is available in our SFR-list. We derive a sponge-like quality of softness by unraveling it and rolling it up and using it instead of introducing a sponge.

5.2 Apply Problem Analogues	Example
Consider solving the problem (using the wording of the physical problem of IFR-2 but this time taking into account the SFR resources that were identified in Part 4) by analogy to similar problems solved in the past.	The chosen IFR-2 in 3.6 was: The operating space <where the mouse is captured> should provide <a trapping arm composed of soft and hard material> by itself during the operating time <during capturing >. Consider problems for which the solution had to provide something that was soft and hard. For example, a motor vehicle has to have a solid steering wheel and controls to be driven, but they need to be soft to protect a person in the event of a crash. An airbag balloon is used to protect the driver. In this case, by analogy

| | we could inflate an overhanging balloon to trap the mouse. |

5.3 Apply the Principles for solving Physical Contradictions	Example
Consider the Inventive Principles in the Table of Specific Inventive Principles to Solve Physical Contradictions (see Appendix 7).	The table is shown in Step 2.1. The user should create ideas from the principles as an exercise.

Note that in ARIZ-85C there is a specific "Table of Transformations" that the user is referred to at Step 5.3. It may be found in the TRIZ body of work. It contains many ideas similar to those captured in the table, how to separate, use of phase changes, etc.

5.4 Apply the directory (database) of Physical Effects	Example
Consider solving the problem using scientific effects. If you do not have access to a database then see Chapter 6. Part 4 of Chapter 6 has the Algorithm for How to Search for Scientific Effects to Solve a Physical Contradiction Problem	To solve the IFR-2 physical contradictions. Search for how to solidify, harden, etc. How to make soft, fluid, etc.

Part 6 offers several ways of reformulating the original problem. Often when first defining a problem it is subject to psychological inertia. After working on it, it may become clear that the problem definition was incorrect.

6.1 Transition form a Physical Solution to a Technical Solution
6.2 Check there is only one problem
6.3 Choose the other technical contradiction
6.4 Reformulate the mini problem

6.1 Transition from a Physical Solution to a Technical Solution	Example
Propose ideas for how to actually apply you solution, to design a way or device for implementing the solution	For the IFR-2 solution: The operating space <where the mouse is trapped> should provide **<a powder (soft) and solid (hard) material of composition of trapping arm>** by itself during the operating time <during trapping>.

It is proposed to attach conductive wires to an electromagnet underneath the frame. The trapping arm, instead of trapping the mouse, actuates the release of a highly flexible silk bag containing ferromagnetic powder. The bag falls on the mouse without harming it. A delay timer also triggered by the trapping arm, it operates a switch to power the electromagnet that solidifies the powder under the influence of a magnetic field and holds the mouse. |

381

6.2 Check there is only one problem	Example
Check the problem formulated in 1.1 is not actually several problems and try to break it down to several sub-problems.	

It is often useful to build a functional model of the problem to try to break the problem down into simpler steps.

6.3 Choose the other technical contradiction	Example
Return to Step 1.4 and choose the other technical contradiction	In the case of the mousetrap we would return to Step 1.4 and choose TC-2 then proceed through ARIZ again.

This means choose the other technical *conflict*. Standard TRIZ and therefore ARIZ-85Cdoes not define conflict and contradiction differently. If you chose TC-1 then go back to Step 1.4. and process TC-2 through subsequent steps of ARIZ.

Note that using the other technical conflict is valid from Step 1.5 up to 3.2. At 3.3 to 3.6 the problem is a physical contradiction, and as we know from Chapter 9 Part 1 "Physical Contradiction - One Physical Conflict" there is only one physical conflict, so users often only repeat only up to 3.2. But at Part 4 we can treat the TC's separately again, Smart Little People is in Part 4 and those are normally drawn for each TC. For Part 5 onwards the problem is formulated as a physical contradiction and treated again as only one problem. We feel this is too complex. At 6.3 we recommend that the user returns to Step 1.4 and repeat every step again up to 6.3 using the other technical conflict that was not chosen at 1.4. Even if a particular step does deal with a physical contradiction, it is seen from a different perspective - the problem is in the opposite state of the contradiction (in our example, low force trapping arm) and it does no harm to repeat the process. More ideas might occur to the user. We recommend 6.3 should read "Return to Step 1.4 and choose the other technical conflict and repeat each step again up to 6.3 if needed."

6.4 Reformulate the mini problem	Example
If there is still no answer, reformulate the mini problem (Step 1.1). State the original mini problem Now consider how the problem can be redefined. It may be useful to re-state the problem at the supersystem level.	Original Mini-problem: It is necessary, with minimal changes to the system to <capture the mouse> and <to not damage the mouse>. Revised Mini problem (at supersystem level) It is necessary, with minimal changes to the system to remove mice from my home and to not damage them.

The reformulated mini problem changes the possible solutions. We could buy a cat which would scare the mice away.

Part 7: REVIEW THE SOLUTION

In Part 7 we review the solution (analyze the removal of the physical contradiction). We check the quality of the obtained solution.

7.1 Check the solution
7.2 Check the problem is solved
7.3 Check the patent database

7.4 Check for sub-problems

7.1 Check the solution	Example
Review the substances and fields you used: Was the problem solved using only available (including derived) resources without introducing any new ones. Can self-controlled substances be used (eliminates the need for further additions). If not go back and try to make the solution more ideal. Can you use derived or modified resources or a "self regulating substance (e.g. a magnet self actuates at the Curie point).	Some solutions used only available resources. Others introduced new resources. Our chosen: Loaded trap Sprung trap The rotating wall trap - solution is from Step 4.3

7.2 Check the problem is solved	Example
a. Did the solution satisfy IFR-1? "The X-resource itself....."	IFR-1 The **direction and shape** of the trapping arm without complicating the system and not causing harmful side effects, eliminates <all damage to the mouse> <during capture> <where the mouse is confined> while preserving the ability to <completely confine the mouse>. Almost, there is some additional cost of materials for making it cylindrical.
b. Does the solution remove the physical contradiction?	Yes, the "arm" which is now the rotating wall is absent with respect to damaging the mouse and present with respect to confining the mouse. The physical contradiction is removed.
c. Does the system use one easily controlled element (part/resource)? Which part and how is it controlled.	Yes, the spring in the center is triggered by the cheese movement.
d. Does it work repeatably over many cycles (is it robust)?	Yes

If the answer to any one of these is no, then go back to 1.1 and start again.

7.3 Check the patent database	Example
Is it infringing a patent or should it be filed as a new patent?	Existing patent

7.4 Check for sub-problems	Example
Are there any new sub-problems (secondary problems) as a result of the solution?	Yes, disposal increases land fill and material usage.
What cost or harmful effects are created by	We plan to develop a re-useable one.

the solution and how will these be addressed?	

It is a good idea to check if the solution to a specific problem can be applied to other similar types of problem. The purpose of Part 8 is to maximize the use of resources. Sometimes a solution can be applied to many similar problems.

8.1 Specify what needs to be changed in the supersystem.
8.2 Can the changed system or supersystem (because of your new system) have new applications?
8.3 Apply the solution concept to other problems.

8.1 Specify what needs to be changed in the supersystem	Example
Changes in the supersystem are:	Changes in the supersystem are: If, for example, the solution in 4.3 will require more display room in shops than existing mousetraps and more landfill space.

8.2 Can the changed system or supersystem (because of your new system) have new applications?	Example
Can your modified system be used in a new way?	Yes, it can be used to keep candy in, a new type of candy box that holds chocolate mice. Once consumed, it can be used as a mousetrap.

8.3 Apply the solution concept to other problems	Example
List other applications for the concept.	The concept can be used as a type of door for separating different environments. The person enters a chamber through the door. Once inside, the walls rotate until the entrance to another room is open.

Each problem solved by ARIZ should increase creativity. Part 9 captures the learning from having applied the ARIZ process for future reference.

9.1 Compare the real process you actually used with the theoretical (ARIZ-85C). Write down any actual differences of the real sequence as key learning for future development of ARIZ
We did not follow exactly the sequence of ARIZ-85C. After Step 1.3 instead of moving to Step 1.4 we tried to solve both technical conflicts by using inventive standards to try to improve the harmful action for both TC-1 and TC-2. Basically we jumped to Step 1.7 and applied inventive standards for both conflicts.
At 3.2 We considered multiple resources not just raw resources.
We created several macro and micro physical contradictions. Since IFR-2 is derived from the macro or micro physical contradictions created in Steps 3.3 and 3.4. from which IFR-2 is sourced.

384

We had no way to know which resource of part of it might form the physical contradiction that best solved the problem, therefore we went back and forth selecting different IRF-2's when applying subsequent tools simply to demonstrate the solution examples.

Although we focused on fixes that require minimal change to the system for IFR-1 and to try to achieve IFR-2 using only available resources to get the operating space itself to achieve the physical contradiction, we did not let the thinking restrict the ideas we created although it helped us not create extremely radical expensive ideas like nuclear radiation or lasers to stun the mouse.

At Step 4.1, instead of drawing Smart Little People for one technical conflict, we drew both, plus the transition. This is because either conflict may have a null positive action which can't be drawn.

Throughout, to solve a physical contradiction we used separation, satisfaction and bypass supported by inventive principles detailed in the "Table of Specific Inventive Principles to Solve Physical Contradictions."

We recommend 6.3 should read "return to Step 1.4 and choose the other technical contradiction and repeat each step again up to 6.3 if needed."

9.2 Compare the solution with the TRIZ knowledge base. Add the solution to Scientific Effects or Inventive Standards

The idea of using a rotating mousetrap as described in Step 4.3 has been added to the knowledge base.

Note that this is an example of how ARIZ-85C can be used. Normally only a few parts of ARIZ are necessary (typically 1 through four).The example responses and solutions given for each step are there to demonstrate to the user by example what response is to be given for each step not how ARIZ is actually used. For example, to demonstrate how a toolset is used, a plumber would show each tool individually, to use the toolset in real life a plumber would only use those needed to complete a specific job. The example is not an attempt to try to create a new type of mousetrap.

END

Index

387

388

Printed in Great Britain
by Amazon.co.uk, Ltd.,
Marston Gate.